STUDENT UNIT GUIDE

NEW EDITION

Edexcel AS Geography Unit 1
Global Challenges

Cameron Dunn and Sue Warn

PHILIP ALLAN

Philip Allan Updates, an imprint of Hodder Education, an Hachette UK company, Market Place, Deddington, Oxfordshire OX15 0SE

Orders
Bookpoint Ltd, 130 Milton Park, Abingdon, Oxfordshire, OX14 4SB
tel: 01235 827827
fax: 01235 400401
e-mail: education@bookpoint.co.uk
Lines are open 9.00 a.m.–5.00 p.m., Monday to Saturday, with a 24-hour message answering service. You can also order through the Philip Allan Updates website: www.philipallan.co.uk

ISBN 978-1-4441-4761-2

Impression number 5 4 3 2 1
Year 2015 2014 2013 2012 2011

Printed in Italy

Hachette UK's policy is to use papers that are natural, renewable and recyclable products and made from wood grown in sustainable forests. The logging and manufacturing processes are expected to conform to the environmental regulations of the country of origin.

P01923

Contents

Content Guidance

Questions & Answers

Getting the most from this book

Questions & Answers

Exam-style questions

Examiner comments on the questions
Tips on what you need to do to gain full marks, indicated by the icon ℮.

Sample student answers
Practise the questions, then look at the student answers that follow each set of questions.

Examiner commentary on sample student answers
Find out how many marks each answer would be awarded in the exam and then read the examiner comments (preceded by the icon ℮) following each student answer. Annotations that link back to points made in the student answers show exactly how and where marks are gained or lost.

Unit 1: Going global   Topic 2

℮ 3/5 marks. Grade C awarded. **a** The first statement is wrong, and suggests the student is not clear about the meaning of adaptation and mitigation. **b c** Two examples of mitigation are given although they are similar, plus the last example **d** which is more detailed and makes reference to carbon emissions. The answer has examples but is unbalanced.

Question 4

℮ These 5-mark questions, which usually appear at the end of a Section A question, often have a 'sting in the tail'. They could:
- use a phrase such as 'costs and benefits' or 'advantages and disadvantages', which requires you to cover both to write a balanced answer and achieve all of the marks
- have a very specific key word such as 'source' or 'host' countries, 'ecological' or 'natural hazards' — if you do not write about the key words in the question your mark will be low
- state 'using examples', which means you should refer to specific named places, players, organisations or schemes. If you don't refer to examples, you will not get full marks

On the move

Explain the costs and benefits of recent eastern European migration for source countries.   (5 marks)

℮ Answers need to cover costs and benefits, but only for the origin countries (sources) of migrants not the destination countries (hosts). The migration type is post-2004 A8 migration from countries such as Poland.

Student answers

Student A
Countries such as Poland have lost up to 1 million workers due to emigration to the UK, **a** possibly meaning worker shortages in Poland due to the Brain Drain effect. **b** Other costs include an ageing population in Poland if workers stay abroad. Benefits include over 6 billion euros sent home in remittances in 2006 and a falling unemployment rate in Poland. Returning migrants could bring back new skills and ideas. **c**

℮ 5/5 marks. Grade A awarded. **a** This is a good answer as it has factual detail about a named A8 country and it shows both **b** costs and **c** benefits, so is balanced. Two benefits plus two costs supported by examples and factual detail ensures maximum marks.

Student B
When people leave they cause a brain drain due to labour leaving the country. **a** Also, there could be fewer people left behind to fill jobs as well as pay taxes, e.g. for benefits. **b**

℮ 2/5 marks. Grade D awarded. This answer is very general, and makes no reference to an A8 source country. **a b** Two costs are outlined; the point about taxes and benefits could have been extended by referring to dependency ratios. This is an unbalanced answer.

About this book

This guide is for students following the Edexcel AS Geography course. It aims to guide you through Unit 1: Global challenges.

The **Content Guidance** section provides a detailed guide to 'World at risk' and 'Going global'.

The **Questions & Answers** section provides examples of Section A and Section B questions with outline mark schemes and student responses at different grades, together with examiner comments.

Overview of Unit 1

Unit 1 examines issues associated with 'World at risk' and 'Going global'. This guide provides you with a summary of the key concepts and structures that you need to study the unit effectively and to revise for the exam.

The headings for each of the 14 sections follow the specification closely. In the specification, the left-hand column is what you need to learn — this is the one the examiners will use when setting questions in the exam. The right-hand column suggests how you might study the concepts and gives you a context for your studies.

As well as using your textbook, you need to keep up to date by following key themes such as Polish migration, global trends in hazards or the latest information on climate change, using articles in newspapers and websites.

While the unit looks at issues on a global scale, there are several opportunities for you to carry out locally based research, such as exploring the roots of your community, assessing local hazard risks or looking at how you and your family could reduce your carbon footprint.

This unit has six compulsory in-depth case studies. These are:
- Global warming in the Arctic (1) and continent of Africa (2)
- Multiple hazard hotspots in the Philippines (3) and California (4)
- Migration flows — eastern Europeans to the UK (5), and UK retirement migration to the Mediterranean (Spain) (6)

Apart from these case studies, which you need to learn in detail, you should learn shorter examples to use in the exam, so that you can support and develop your arguments. It is also well worth building up a vocabulary list of all the key words you learn over the course to improve your understanding of the concepts.

Examiner tip
Learn these compulsory case studies as you will be able to use them in Section A or B questions. Look at the specification to see the precise focus.

Content guidance

World at risk

'World at risk' contains two interlinked strands — **global hazards** and **climate change** — which together increase the risk from disasters, especially for the world's most vulnerable people.

Figure 1 Topic concept map

Global hazards, which can be classified into two major categories, **hydro-meteorological** and **geophysical**, cause increasing numbers of disasters around the world. Frequently a number of hazards occur together (i.e. are **spatially concentrated**). Where the people are already vulnerable, because of poverty and high population densities, **multiple-hazard hotspots** occur. A study of **global trends** confirms that the frequency and magnitude of hydro-meteorological disasters is increasing as a result of a possible link to **climate change**.

Climate change is considered by many to be the world's greatest problem. Technically, it is a **context hazard** and therefore a 'chronic global-scale threat' to both the environment and people.

This topic explores the causes and impacts of climate change and solutions to the problem. It considers whether recent global warming is unprecedented in the history of climate change and is therefore **anthropogenically** caused. Short-term climate change encompasses not only global warming but also the impact of **oscillations** such as El Niño.

Examiner tip

The diagram shows how global hazards and climate change are interrelated, so think about the links as you are revising.

Global hazards

Risk and vulnerability in an unequal world

- **Environmental hazards** can be defined as 'the threat posed to humans and the environment by natural events originating in it'. These natural events can be classified in many ways, but it is useful to categorise them as geophysical or hydro-meteorological. Humans invariably make the event worse and there are very few hazards that are caused by purely physical factors.
- **Geophysical hazards** are formed by tectonic/geological events (earthquakes, volcanoes and tsunamis).
- **Hydro-meteorological hazards** are formed by hydrological (floods) and atmospheric (storms and droughts) processes.
- **Context hazards** are widespread threats arising from global environmental changes such as climate change or from a major hazard such as a super-volcano.
- **Disasters** are the 'realisation' of hazards to cause social impacts such as loss of lives and livelihoods, and economic impacts such as damage to goods and property (i.e. the actual *consequences* of a hazard).
- **Risk** is the probability of a hazard occurring and becoming a disaster as a result of deaths and loss of livelihoods, goods and property (i.e. the likely *exposure* to a hazardous event).
- **Vulnerability** is shaped by the underlying state of human development, including inequalities of income, opportunity and political power, all of which marginalise the poor. Poverty-stricken people frequently depend directly on the land for food (subsistence), live in fragile eco-environments and have poor health and well-being. The result is that high risk is combined with an inability to cope with the adverse impact of natural hazards and climate change (at all levels — individuals, communities and countries). These relationships are summarised in Figure 2.

Knowledge check 1

Do you understand the difference between a hazard and a disaster?

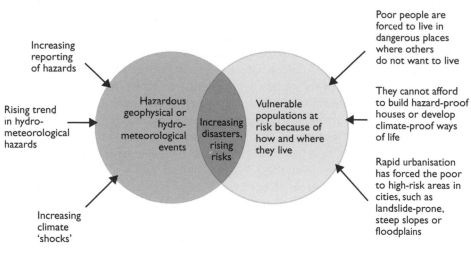

Figure 2 Degg's model

Risks and vulnerability are now greater as a result of climate change, with more hydro-meteorological disaster events, which add to the pressure on countries that have a limited ability to adapt and where coping strategies are already stretched.

Examiner tip
Learn the risk equation and understand why the risk is getting worse.

Rising risks

The **risk equation** measures the level of hazard for an area:

$$\text{risk } (R) = \frac{\text{frequency or magnitude of hazard } (H) \times \text{level of vulnerability } (V)}{\text{capacity of population to cope and adapt } (C)}$$

$$R = \frac{H \times V}{C}$$

The risk is getting worse for many communities and countries because:
- the frequency and magnitude of hazards are increasing with climate change
- vulnerability is increasing as a result of unsustainable development leading to poor land use and environmental degradation
- the capacity to cope is decreasing owing to poverty and urbanisation

Note that the risk is much lower in economically developed countries. These countries have the resources and technology to provide protection from the worst impacts of disasters.

Examiner tip
Be careful not to blame every extreme weather event on global warming. Atmospheric/oceanic oscillations are a more likely culprit. Global warming tends to affect short-term climate trends.

Recent climate change: the world's greatest hazard?

It is important to use the correct terminology — see the 'Climate change survival guide' below. Climate change can be classified using timescales (long, medium and short term). Short-term or recent climate change, most marked since the 1960s, is not only the result of global warming but also of the increasing impact of atmospheric/oceanic oscillations such as the El Niño/La Niña effect.

Climate change survival guide
- **Climate change** is any marked trend or shift in climate (average weather over 30 years) that shows a sustained change in the average value for any particular climatic element (rainfall, drought, storminess etc.) Clearly, for longer- and medium-term climate change, evidence other than records of weather measurement has to be used.
- **Climate variability** refers to the differences in climate from one year to another. More extreme weather seems to be associated with the current climate.
- **Global warming** refers to a consistent recently measured rise in the average surface temperature of the planet (i.e. at the global scale). However, note that neither the rise in temperature nor the related impacts on atmospheric and oceanic circulation are likely to occur uniformly across the globe.
- The **greenhouse effect** is a natural process that warms the Earth's atmosphere due to the trapping of heat that would otherwise be radiated back into space — it enables the survival of life on Earth.
- An **enhanced greenhouse effect** occurs when the amount of greenhouse gases (the main ones are water vapour, carbon dioxide, methane, nitrogen oxide and CFCs) increases. Scientists argue that this is almost certainly the result of anthropogenic (human) activity such as the burning of fossil fuels (oil, coal, gas) and it leads to global warming.
- **Fossil fuels** are rich in carbon and burning them releases carbon dioxide.
- **The tipping point** is the point at which a system switches from one state to another. In terms of climate change, scientists have identified this tipping point as a 2–3°C rise in temperature. This small rise in temperature could be enough

to lead to dramatic and possibly cataclysmic changes in the environment which are probably irreversible. For example, the failure of the Gulf Stream to reach northwest Europe would contribute to a much more hazardous world.

- **The 'hockey stick' graph** shows the average global temperature change going back 1,000 years. It is so called because it shows a long flat period followed by a rapid increase in temperature in the last 50 years.
- **A feedback mechanism** is a process whereby the output of a system acts to amplify (positive) or reduce (negative) further output. For example, the melting of Arctic permafrost would lead to the release of large quantities of trapped methane, which would itself lead to further global warming.
- **Global dimming** is the phenomenon whereby particles in the atmosphere, largely from industrial pollution, have led to less solar radiation reaching the Earth (in fact this may have masked the initial 1960s global warming).
- **IPCC** is the **Intergovernmental Panel on Climate Change**, which was set up in 1988. It is the central scientific voice on the issue.

> **Knowledge check 2**
>
> Is the melting of Arctic permafrost a positive or a negative feedback mechanism?

Ten reasons why recent climate change might be the world's greatest hazard

1 Climate change is a **global problem**: all areas of the world will be affected to a lesser or greater degree. However, the impacts are various, and often unique to a particular country.

2 Climate change is a **chronic (ongoing) hazard**. It will have an enormous range of direct impacts linked to rising temperatures, which can themselves amplify other problems. These include climate belt migration towards the poles, which will have a severe impact on ecology and wildlife and lead to the spread of diseases such as malaria. Rising temperatures also have an impact on both atmospheric and oceanic circulations. Rising ocean temperatures may be a cause of more frequent atmospheric/oceanic oscillations (such as El Niño) and increasing frequency and magnitude of hurricanes.

3 Climate change is not just an environmental problem. It has a **range of impacts** on societies and economies, threatening their very existence in terms of water availability, food security and health and wellbeing.

4 Climate change is **unpredictable**. It could become catastrophic depending on whether, or how fast, the tipping point is reached. This depends on both physical and human factors:
   - There are a number of interlocking systems that can affect each other via **positive** and **negative feedback**. For example, rising temperatures melt glaciers and ice sheets. This leads to more incoming solar radiation being absorbed by land and sea, because ice and snow have a greater albedo (reflective power) than rock and water. Thus melting of the ice sheet leads to yet more melting — an example of positive feedback. The melting ice sheets could lead to the ocean becoming diluted, which could have an effect on ocean current circulation, thus weakening the warming powers of the North Atlantic Drift (negative feedback).
   - The human factor is related to how effectively the world can **mitigate** the damaging effects of climate change by reducing the production of greenhouse gases to slow down the rate of global warming. Many people argue that only with radical solutions will the impacts of climate change be halted.

5 Climate change has many **indirect impacts** too. For example, thermal expansion of the oceans as water expands with heating leads to rising sea levels, a process

that may reach catastrophic levels if the large polar ice sheets (Arctic and Antarctica) melt. Rising sea levels have already created the world's first 'environmental refugees' in the islands of the Ganges delta and the South Pacific.

6 Climate change prediction requires successful **modelling** of the future by scientists in the IPCC for both direct impacts (rising temperatures, more extreme weather) and indirect impacts (rising sea levels, changing ecosystems). The sheer complexity of the calculations makes this process extremely difficult, with huge variations in the scenarios presented. This makes it particularly hard to convince people about the severity of climate change.

7 It is difficult for scientists and researchers to **separate the effects** of global warming from those of other influences on present-day weather such as the El Niño/La Niña cycle. Many of the changing weather patterns — such as more extremes with greater climate variability, or increasing frequency of droughts, big storms and widespread flooding — may be attributable to global warming, but are more likely to be linked to atmospheric/oceanic oscillations. The droughts and floods in 2007 are an example.

8 Climate change requires **global solutions**, primarily in tackling carbon emissions. However, there are huge political problems. It is unjust to have a two-speed, unequal world in which more developed countries pollute and less developed countries are vulnerable victims. A further problem is the emergence of China, India and Brazil as rapidly expanding economies, and a resurgent Russia, which feel that their industrial development should not be jeopardised by targets for reducing greenhouse gas emissions.

9 Until recently the issue of climate change was **strongly contested**. While there is now a broad consensus among scientists, led by the IPCC, that levels of greenhouse gases and global temperatures are rising and that there is a link between them, there are still sceptics. There are two such groups: those who find weaknesses in the data and their interpretation; and those with vested interests (for example, oil companies), which claim that global warming and the resulting hazards are due to factors other than the burning of fossil fuels. However, the awarding of the 2007 Nobel peace prize to Al Gore (who presented the film on global warming *An Inconvenient Truth*) and to the Chair of the IPCC gave legitimacy to the campaign to take action on climate change.

10 Climate change is a problem of enormous scale and proportions, and so is the world's most **costly problem to solve**.

Global hazard trends

Is the world becoming more hazardous?

The issue

To find out whether the world is becoming more hazardous or not, you need to use statistics from key websites (see below) to discover whether more hazard events are occurring and, if so, of what type and why. Remember that many textbooks use the terms 'hazard' and 'disaster' interchangeably. There may not be more hazards of all types occurring, but a higher percentage are becoming disasters because they affect increasing numbers of vulnerable people. There is also the fact that hazards and disasters are hot news, and scientists, engineers, governments, insurance companies and risk managers all take great interest in them. It may be that there are not actually more hazards, but more are being reported because of this media interest and the improvements in information and communications technologies.

Disaster statistics

Disaster statistics come from a number of sources. Governments report disaster statistics to UN agencies, e.g. WHO (**www.unisdr.org/disaster-statistics/introduction.htm**). These quasi-official statistics are supplemented and cross-checked by designated groups that monitor media and internet reports and also field data from NGOs working on the front line.

The pre-eminent authority is CRED (Centre for Research on the Epidemiology of Disasters, **www.cred.be**), supported by the World Bank. It administers the key database: **www.emdat.be/**. Both Munich Re (**www.munichre.com**) and Swiss Re (**www.swissre.com**), which are re-insurance multinationals, provide disaster databases for their own internal use.

Trends in the number of reported disasters

- The number of reported natural disasters has risen sharply since the 1960s.
- This trend is particularly marked for **hydro-meteorological disasters**. The increase in floods and storms, and more recently in droughts, is responsible for driving the overall trends. Floods and storms together accounted for 70% of all disasters globally in the last 40 years.
- **Geophysical disasters** such as earthquakes and volcanoes have increased slightly since the 1970s, possibly as a result of more reporting of disasters and also because of the rising number of vulnerable populations living in hazard-prone areas. However, in general, the trend fluctuates. Indonesia is currently a 'hotspot' for tectonic activity, and it has two very active volcanoes. It has experienced five major earthquakes and several tsunamis since 2005. All of these events are the result of a highly mobile plate boundary where the Indian plate is being subducted beneath the Burma plate (at a destructive boundary).
- **Biological hazards** (bio-hazards) have also shown an upward trend from a low base, often numerically exceeding geophysical hazards. This may be linked to global warming which leads to more pests and diseases.

Examiner tip
It is up to you to use the various databases and to explore the evidence. Research the databases to get the latest year's statistics as credit will be given for up-to-date research.

Examiner tip
Note how the rising trend fluctuates and some years such as 2004/5 or 2009/10 seem to be more hazardous. Try to work out why — is it the oscillations or is it to do with where the hazards occurred?

Knowledge check 5
Which type of hazard is driving the overall rising trends in disasters?

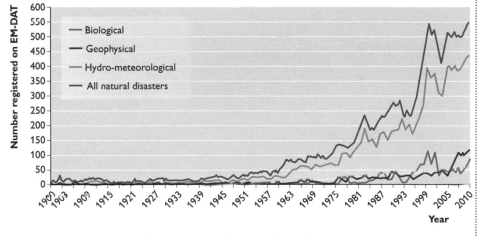

Figure 3 Hazard trends, 1900–2005

The reasons for the rising trend in hydro-meteorological disasters are complex and include both physical and human factors.

The media almost always state that the rising trend is a result of global warming, suggesting for example that rising temperatures lead to warmer oceans, thus 'spawning' more hurricanes, and cause stronger convectional cells — again creating more hurricanes. Equally, rising temperatures increase evaporation, which in turn contributes increasing rainfall and therefore more flooding.

The IPCC suggests that a more extreme climate (resulting from global warming) will increase the risk from certain hazards. In some cases, it may lead to more severe events (greater magnitude); in other cases, the events may occur more often (higher frequency).

Of equal importance are oscillations such as the El Niño/La Niña cycle. The cycle, which affects winds and ocean currents, lasts for 7–10 years, in which 1–2 years are El Niño years and 1–2 years are La Niña years. The cycle has a variety of effects leading to hazardous weather in many parts of the world (these links are known as teleconnections):

- During El Niño years there is unusually heavy rain in Peru and California, and dry conditions in Indonesia and Australia (2006 drought).
- During La Niña years there is unusually heavy rain in Australia and Indonesia, and drought in the Americas (Californian drought and wildfires) and various parts of Africa.
- Hurricanes in the Atlantic are usually increased in La Niña years and reduced in El Niño years.

Some scientists even suggest that global warming may increase the frequency and intensity of these oscillations.

Purely human factors also play a significant part in rising trends in disasters. Figure 4 summarises these factors.

Examiner tip

Get up to date: 2009 was a weak El Niño year and 2010 was a very strong La Niña year. Research the impacts such as floods in Pakistan and eastern Australia.

Knowledge check 6

Which type of hazards are likely to be a particular danger at the coast?

Examiner tip

Think about your compulsory case studies of California and the Philippines. Work through the factors shown and consider whether they have led to increased disaster frequency.

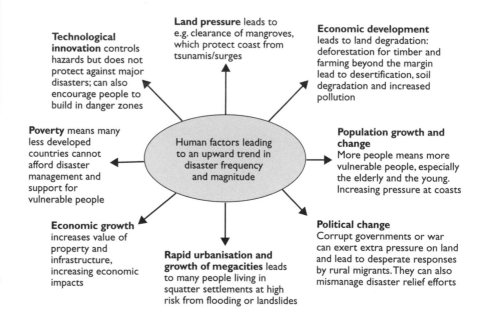

Figure 4 Human factors increasing disasters

Trends in the social and economic impacts of disasters

The statement 'Deaths are falling, whereas the number of people affected by disasters is rising and economic losses are escalating' summarises the situation globally. However, economic development also plays a part. A general statement such as 'Less developed countries experience more deaths (social costs) and more developed countries experience more damage (economic costs)' has some truth.

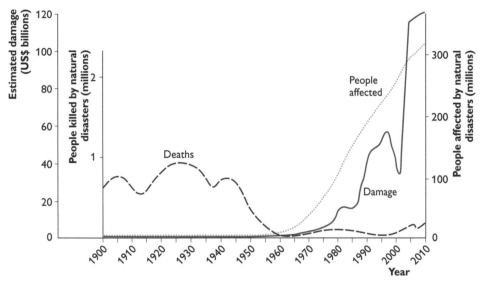

Figure 5 Global trends in deaths, people affected and economic losses from natural disasters, 1900–2006

Examiner tip

This is just the sort of diagram you should learn for a Section B answer. Annotate the trends to add explanation.

Knowledge check 7

What is the difference between an economic and a social impact?

Falling deaths?

As Figure 5 shows, the number of people reported killed by natural disasters dropped dramatically before the 1970s, but it has levelled off in recent years to a fluctuating trend, with a rise in 2004–05 (mega events). This is largely the result of better disaster management before, during and immediately after the disaster:

- **Before a disaster** there is better mitigation (use of hazard-proof/climate-proof structures) and improved disaster preparedness using education, community training and high-technology warning systems.
- **During the disaster** — in many cases emergency responses have improved, with local participation in effective disaster relief programmes (e.g. Hurricane Sidr in Bangladesh in 2007).
- **Post-disaster** recovery programmes are also better coordinated, often using GIS to pinpoint problems. Studies of the Kashmir earthquake (2005), Hurricane Katrina (2005), Typhoon Nargis (2008) and the south Asian tsunami (2004) do however highlight many problems, but there is no doubt that, especially in developed areas, deaths are minimised by an effective disaster response. Following the Sichuan earthquake in 2008 the Chinese government attempted a prompt disaster response in spite of numerous difficulties.

All this is against a backdrop of increasing hazards and disasters, especially from hydro-meteorological causes.

Knowledge check 8

Why are there occasional 'blips' in the downward trend in deaths, e.g. 2004/5, 2009/10?

Rising numbers of people affected

The definition of 'being affected' includes loss of home, crops, animals, livelihood, and often a decline in health/quality of life for a designated period (usually 3 months).

As Figure 5 shows, in the 1970s on average 100 million people a year were affected by disasters, but in the early twenty-first century this has risen to over 300 million a year (with peaks up to nearly 700 million as a result of the 2004 tsunami and other mega disasters — not shown on the graph). Increased vulnerability due to a less favourable risk equation (see page 8) is responsible for this rising trend, as some 80% of the millions affected are concentrated in areas of low and low-to-medium human development. Countries with the greatest numbers of people affected are almost exclusively less developed countries and new industrialised countries (NICs) such as India and China.

Economic losses from disasters

Economic losses from disasters are growing at a faster rate than the number of disasters. The value of the losses is escalating compared with the insured risks, because hazard-prone areas in general have high insurance premiums, and some businesses and many poor people (who are forced to live in hazard-prone areas) cannot afford or even access insurance. Rising numbers of disasters are occurring in less developed countries where insurance is rarely an option. Mega-disasters such as Hurricanes Andrew and Katrina can lead to huge rises in economic losses for a single year, resulting in a fluctuating but nevertheless overall rising trend.

There is a tendency to over-emphasise the economic damage and losses experienced in more developed countries. In absolute terms, there is no doubt that because of the value of possessions, the sophistication of installations and infrastructure, and the scale of domestic insurance claims, the amounts involved in developed countries are large.

However, in relative terms, economic damage from natural disasters tends to be higher in less developed countries, mainly because of their high dependence on one or two economic sectors, such as tourism or cash cropping of bananas. The damage forms a very high percentage of their very low annual gross domestic product (GDP). For example, Hurricane Mitch in 1998 caused a 50% loss in GDP for Nicaragua. By contrast, Munich Re estimated that between 1994 and 2003 Japan suffered $166 billion of economic damage from national disasters but this was only 2.6% of its GDP.

Another reason why the rate of economic losses is increasing faster than the occurrence of disasters is the growing economies of NICs such as India, China and the 'Tiger' economies of Korea, Taiwan, Singapore and Malaysia.

Global hazard patterns
Exploring local hazards and links to climate change

You must research hazard risks in your local area or another small-scale area. A starting point is your local reference library. Research local records including newspaper archives and websites that record historic and present-day hazard events. You will probably find rising numbers of floods and storms, but less snowfall.

Knowledge check 9

Explain why economic losses from disasters are growing at a faster rate than the number of disasters.

Knowledge check 10

What is the difference in calculating the impact of a disaster in absolute and relative terms?

Examiner tip

Do try to research your local area. Most areas in the UK have experienced increases in extreme weather (including storms, gales and drought) and both winter and summer floods with landsliding – all possibly related to short-term climate change. Some areas have even experienced earthquakes along old fault lines.

When you are researching climate change you will find it is difficult to get **climatological data** on changing climates for your local area, but there are a number of websites that provide this information at a regional or national scale. The Climate Research Unit (**www.cru.uea.ac.uk**) and the Tyndall Centre (**www.tyndall.ac.uk**) provide useful maps and data for the UK from which you can extract information for your local area.

Another interesting method of research is the science of **phenology**, which looks at key indicators to measure the response to climate change. Figure 6 summarises some of these indicators. All the results seem to indicate that in recent years (especially 2007) we have had warmer winters and earlier springs. Natural events such as those shown may have moved up to 15 days earlier over the last 30 years. There are a number of useful websites for this type of research (e.g. **www.naturescalendar. org.uk**, **www.cru.uea.ac.uk/cru/info/iccuk** and **www.ukcip.org.uk**).

Examiner tip

When looking at evidence for global warming it is very easy to get confused. In the UK, January and December 2010 were both very severe months, masking the overall warming trend.

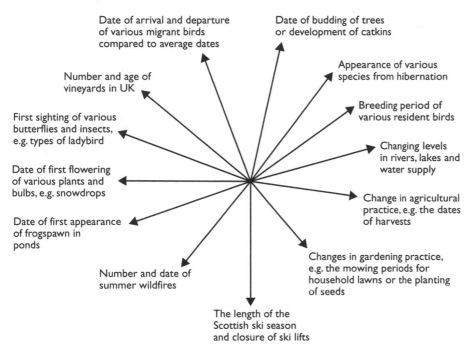

Date of arrival and departure of various migrant birds compared to average dates

Date of budding of trees or development of catkins

Appearance of various species from hibernation

Number and age of vineyards in UK

Breeding period of various resident birds

First sighting of various butterflies and insects, e.g. types of ladybird

Changing levels in rivers, lakes and water supply

Date of first flowering of various plants and bulbs, e.g. snowdrops

Change in agricultural practice, e.g. the dates of harvests

Date of first appearance of frogspawn in ponds

Changes in gardening practice, e.g. the mowing periods for household lawns or the planting of seeds

Number and date of summer wildfires

The length of the Scottish ski season and closure of ski lifts

Figure 6 Possible climate change response indicators

The distribution of the world's major natural hazards

For each of the major hazards (earthquakes, volcanoes, slides, drought, floods and storms), you need to know the following:

- The **distribution** of the occurrence for each hazard. A world map to show this is crucial, as the distribution for each hazard is very different.
- A clear **definition** of each of the six types.
- A basic understanding of the **causes** of each hazard. For example, for geophysical hazards you need to know how plate tectonics affects the distribution of earthquakes and volcanoes.
- Some understanding of the likely **impacts** of each hazard, so that you can weigh up its contribution to disaster statistics (such as deaths and damage) and also see how it affects your case studies of disaster hotspots.

Knowledge check 11

Which of the six hazards listed (a) kills the most people each year; (b) kills the fewest people each year; (c) is known as the silent killer?

The distribution of earthquakes

As can be seen from Figure 7 the main earthquake zones are clustered along plate boundaries. The most powerful earthquakes are associated with destructive or conservative boundaries. Earthquakes are a common hazard and can develop into major disasters when they are of high magnitude (i.e. Richter scale 6+) and occur in a densely populated/vulnerable area (e.g. Kashmir, 2005). Earthquakes not only cause primary hazards from ground movement and ground shaking, but also cause secondary hazards such as landslides and even tsunamis (e.g. south Asian tsunami, 2004).

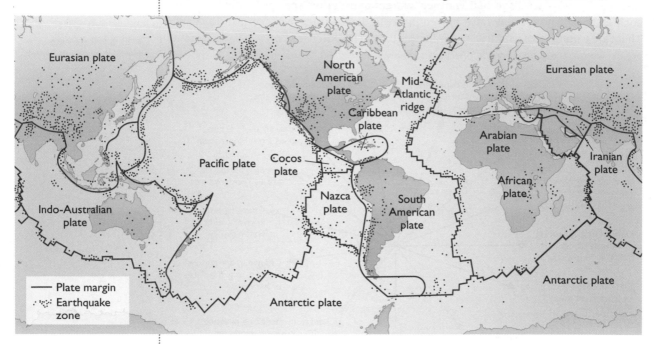

Figure 7 Distribution of earthquakes

Destructive plate boundaries

Destructive boundaries, where oceanic crust is being subducted beneath a continental plate (e.g. Philippines case study), or where two oceanic or continental plates collide, produce shallow, intermediate and deep earthquakes.

Constructive plate boundaries

Constructive boundaries (where oceanic plates are moving apart) are associated with large numbers of shallow, low-magnitude earthquakes. Most are submarine (under the sea).

Conservative plate boundaries

Conservative boundaries, where the plates slide past each other, produce frequent, shallow earthquakes, sometimes of high magnitude (e.g. along the San Andreas fault system of the western USA).

Other earthquakes

A small minority of earthquakes occur within plates, usually involving the reactivation of old fault lines (e.g. the Church Stretton fault in Shropshire). Occasionally earthquakes

Knowledge check 12

Which type of plate boundary (and why) is not associated with severe earthquakes?

can result from human actions, such as dam or reservoir building, which increase the weight and therefore stress on the land (e.g. in Killari, northern India, in 1993, an earthquake caused by a dam killed 10,000 people).

The distribution of volcanoes

Figure 8 shows the global distribution of active volcanoes. The type of tectonic situation determines the composition of the magma and therefore the degree of explosivity of the eruption, which is a key factor in the degree of hazard risk.

Constructive plate boundaries

Most of the magma that reaches the Earth's surface wells up as volcanoes at ocean ridges, such as the mid-Atlantic ridge. These volcanoes are mostly on the sea floor and do not represent a major hazard to people except where they emerge above sea level to form islands (e.g. Iceland). The east African rift valley, which is a constructive plate boundary, has a line of 14 active volcanoes, some of which can produce dangerous eruptions (e.g. Mt Nyragongo in the Democratic Republic of Congo, 2002).

Destructive plate boundaries

Some 80% of the world's most active volcanoes occur along destructive boundaries. When oceanic plates are subducted beneath continental plates, explosive volcanoes, such as Mt St Helens, are formed. The 'ring of fire' around the Pacific has many such volcanoes, including those in the Philippines.

Knowledge check 13

Which type of plate boundary is not associated with volcanicity and why?

| — Plate boundary | | | | | |
● Active volcano					
1 Azores	6 Galunggung	11 Krakatoa	16 Mt St Helens	21 Popocatapetl	26 Tambora
2 Bardarbunga	7 Grímsvötn	12 Mauna Loa	17 Nevado del Ruiz	22 Redoubt	27 Tristan da Cuhna
3 Cotopaxi	8 Haeimaey	13 Soufrière Hills	18 Nyos	23 Ruapehu	28 Unzen
4 Etna	9 Katmai	14 Mt Pelée	19 Parícutin	24 Surtsey	29 Vesuvius
5 Fujiyama	10 Kilauea	15 Mt Rainier	20 Pinatubo	25 Taal	30 Meru

Figure 8 Global distribution of active volcanoes

Knowledge check 14

Name two areas with active volcanoes that are associated with plumes from hotspots and not plate boundaries.

Knowledge check 15

Explain why the Philippines is such a major landslide area.

Knowledge check 16

What is a drought?

Examiner tip

Describing the distribution of the six major hazards requires some very precise factual knowledge and specialist terminology. Spend time on learning this terminology so that your answer stands out in terms of quality.

Hotspots

Hotspots are localised areas of the Earth's crust (lithosphere) which have an unusually high heat flow, and where magma rises to the surface as a plume (e.g. Hawaii). As a lithospheric plate moves over the hotspot, a chain of volcanoes is created.

Volcanic hazards

Apart from the local impacts of lava flows the most catastrophic impacts of volcanoes are pyroclastic flows, ash falls, tsunamis and mudflows (e.g. in 1983 mudflows from Nevado del Ruiz killed over 22,000 people).

The distribution of slides

Slides include a variety of mass movements, such as rock slides, debris flows, snow avalanches, and rainfall- and earthquake-induced slides.

Landslides

Landslides are the seventh biggest killer with over 1,400 deaths per year, ranking above both volcanoes and drought. Most landslides occur in mountainous areas, often after abnormally heavy rain and/or seismic activity. Human factors also play a part. Deforestation of hillsides in southeast Asia and building on hillslopes in Hong Kong have both led to widespread slides following rain.

Snow avalanches

Snow avalanches are concentrated in high mountainous areas such as the Southern Alps of New Zealand or the Rockies of North America. Avalanches tend to occur on slopes steeper than 35°. An average of 40 deaths a year in Europe and over 100 in North America are caused by avalanches. Recent research has suggested that global warming may be increasing avalanche occurrence, although trends in deaths have slowed because of effective management.

The distribution of drought

Drought has a dispersed pattern — over one-third of the world's land surface has some level of drought exposure (see Figure 9). This includes 70% of the world's people and agricultural value, which means that drought has an effect on global food security.

The causes of drought include the following:

- Variations in the movement of the inter-tropical convergence zone (ITCZ). As the ITCZ moves north and south through Africa it brings a band of seasonal rain. In some years, high-pressure zones expand and block the rain-bearing winds. In southern Ethiopia and Somalia, where farmers depend for food on rain-fed agriculture, famines may result if the summer rains never arrive.
- El Niño can bring major changes to rainfall patterns. In particular (as in 2006), it can bring drought conditions to Indonesia and Australia.
- Changes in mid-latitude depression tracks. In temperate regions, depressions bring large amounts of rainfall. However, if blocking anticyclones form and persist, depressions are forced to track further north, leading to very dry conditions. Droughts in the UK and France (1976, 1989–92, 1995, 2003 and 2006) as well as in the US midwest in the 1930s were all related to this cause.

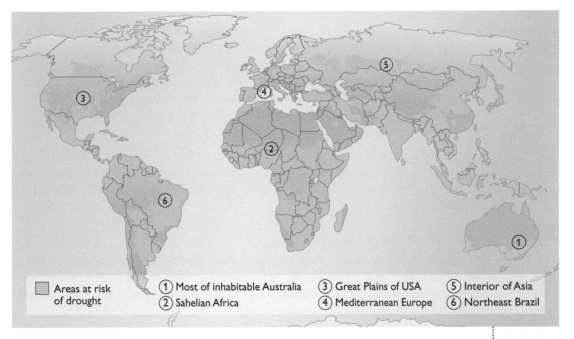

Figure 9 Global distribution of drought-risk areas

The distribution of flooding

Flooding is a frequent hazard and is evident in some 33% of the world's area, which is inhabited by over 80% of its population (Figure 10). Regional-scale, high-magnitude floods are frequent events in India/Bangladesh and China.

Knowledge check 17

How does 'continentality' contribute to drought?

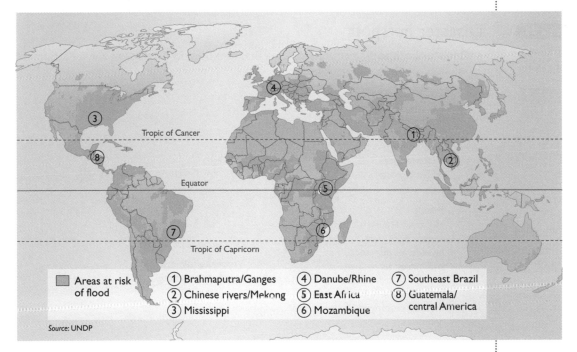

Figure 10 Global distribution of flood-risk areas

The causes of flooding include the following:

- By far the most common cause is excessive rainfall related to atmospheric processes, including monsoon rainfall and cyclones. In temperate climates, a series of depressions sometimes brings prolonged heavy rainfall.
- Intense rainfall sometimes associated with thunderstorms can lead to localised flash flooding. These sudden floods can have a devastating effect.
- El Niño can bring devastating floods, as in Mozambique in 1997 and 2006.
- Rapid snowmelt can add water to an already swollen river system.

In developing countries flooding may lead to deaths by drowning and disease, destruction of food crops and infrastructure and loss of homes. In developed countries it disrupts transport and infrastructure, damages livelihoods and creates high insurance costs.

The distribution of storms

Storms include tropical cyclones, mid-latitude storms and tornadoes. Tropical cyclones (hurricanes in the Atlantic) are violent storms between 200 km and 700 km in diameter. They occur in the latitudes 5–20° north and south of the equator. Once generated, cyclones tend to move westward.

Tropical cyclones or hurricanes will only occur over warm ocean (over 26°C) of at least 70 m depth and at least 5°N or 5°S of the equator in order that the Coriolis effect (very weak at the equator) can bring about rotation of air.

Storms cause damage in several ways, including heavy rain (leading to floods and mudslides), high wind velocity and very low central pressure (leading to storm surges and coastal flooding). They can be devastating in low-lying mega-delta regions (e.g. Hurricane Katrina and Cyclone Nargis).

Disaster hotspots

A **disaster hotspot** is a country or area that is extremely disaster-prone for a number of reasons, as shown on Figure 11.

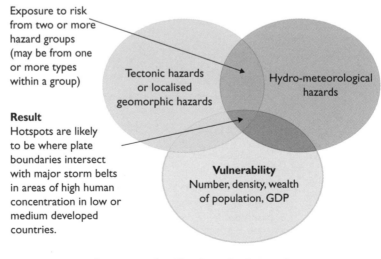

Figure 11 Identification of a disaster hotspot

Edexcel AS Geography

You need to learn the Philippines and California coast case studies in depth with detailed statistics. Learn how to draw a simple sketch map on which you can annotate key facts about the occurrence and impacts of the range of disasters they experience. For each one, think about the types of hazard and why they are concentrated there (with examples of recent events and the deaths and damage caused). Consider how vulnerability can lead to disasters, and the adaptive capacity of the people and government. The interaction between people and the natural events is the key to understanding their impact.

These two case-study areas have been selected because of the contrast in the types of hazard they experience and their states of development.

Table 1 summarises the contrasts between the two areas.

Examiner tip

Learn these in-depth case studies very precisely so that if required you could compare them in terms of hazard types and other factors.

Table 1 The Philippines and California coast compared

	Philippines	California coast
Status	Lower-middle income country	Wealthiest state of the USA
Population	91 million (density 240 per km^2, 2,000 per km^2 in Manila)	40 million (2008)
GDP per capita	$5,000	$40,000
HDI	0.75	0.95 (nearly the highest)
Landscape	Mountainous country, with crowded coastal lowlands. Consists of 7,000 islands, many very small, spread over latitudes between 5°N and 20°N (typhoon belt)	Coastal area contains the huge conurbations of Los Angeles, San Francisco and San Diego
Summary	Classic hotspot where typhoon belt interacts with mobile plate boundary in a rapidly developing country; El Niño/La Niña cycles increase range of hazards	Elongated area ranging from subtropical in north to tropical in south; major tectonic hotspot, which is also subject to extreme weather hazards brought by El Niño/La Niña oscillations

Compulsory case study **The Philippines**

- Typhoons are the main hazard (20 per year).
- Volcanic eruptions are explosive with dangerous lahars (mudflows).
- Earthquakes are common (100% of country at risk).
- Occasional droughts associated with El Niño years.
- Landslides are common in mountain areas.
- The Philippines is a densely populated, rapidly developing country.
- Vulnerability is increased by poverty, deforestation, poor land management and rapid urbanisation.

Advice on drawing your Philippines map

- Practise drawing the outline of the islands (hard!)
- Add the earthquake risk (whole country, but concentrated near plate boundary).
- Indicate areas at greatest risk from typhoons. Essentially the south is less at risk and the northeast at greatest risk. Note the season/months.
- Mark on and name key volcanoes, e.g. Pinatubo, Taal and Mayon (17 are active).
- Mark on Manila (megacity).
- Annotate central highland area as a site of landslides, e.g. Leyte.
- Annotate small islands as sites of local tsunami risk, e.g. Biltran.

Knowledge check 20

Why is the greatest risk from typhoons found in the northern Philippines?

Knowledge check 21

Describe the type of plate movement experienced in the Philippines and explain how it leads to volcanoes.

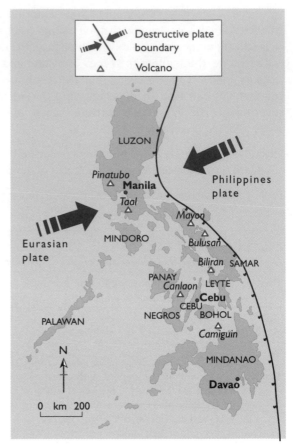

Figure 12 Sketch map of the Philippines

Table 2 shows the types of disasters the country experiences.

Table 2 Disasters in the Philippines, 1905–2007

	Number of events	Total killed	Total injured	Homeless total	Affected total	Damage ($m)	Example
Drought	6	8	0	0	6 million	64,000	April 1998, 2.5 million severely affected
Earthquake	21	9,580	13,051	3,985	2.25 million	844,485	Manila (1990) 6,000 killed
Flood	72	2,716	570	500,000	11.25 million	446,361	July 1972, 2.7 million affected
Slide	25	2,604	381	23,000	310,663	12,258	February 2006, 1,126 killed
Volcano	20	2,996	1,188	79,000	1.5 million	23,961	Taal (July 1911), 1,335 killed Pinatubo (July 1991), 700 killed
Tsunami	5	69	0	0	5,250	6,000	Worst tsunami in 1976
Typhoon	241	35,983	29,178	6.25 million	86 million	9,018,574	November 1991, 6,000 killed

Compulsory case study **The California coast**

Figure 13 summarises the risks and the links between them.

- Earthquakes are the main risk. Large shallow earthquakes occur along the swarm of faults associated with the San Andreas fault — a conservative plate boundary.

- River floods occur in El Niño years, and droughts and wildfires in La Niña years.
- Fogs occur in the San Francisco Bay area.
- Landslides are a frequent secondary hazard from floods or earthquakes.
- Coping capacity is high in this wealthy area, with much high-tech disaster preparedness, and state-wide drills for 'the big one'.
- Economic costs of major disasters are high, especially if disaster strikes in a megacity.

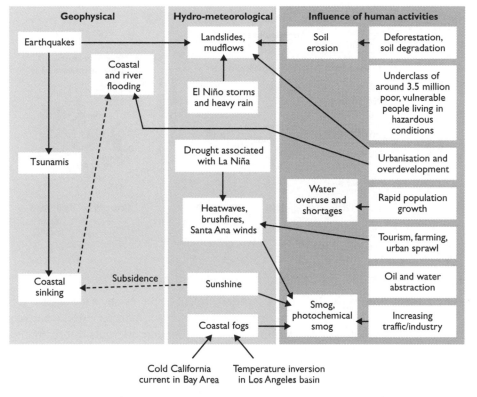

Figure 13 Disaster web for the California coast

Points to consider about multiple hazard zones

What is the effect of being a multiple hazard/disaster hotspot?

1 The hazards often occur as part of a complex web, making their impact greater and more complex to manage.
2 The frequency of events does not always make the impact easier to manage, although in wealthy countries systems can be put in place (contrast California with the Philippines, which only has good systems for geophysical hazards).
3 The magnitude of the hazard event, and the human geography of the area in which it occurs, are still probably the most important factors.

Climate change and its causes

Putting short-term climate change (global warming) in context

Climate change occurs at a range of different timescales, as shown in Figure 14.

Which type of hazard do you consider poses the greatest risk for California?

Which type of disaster does 'the big one' refer to?

Examiner tip

If there is a recent hazard event within your case study area make sure you research it in order to add topicality to your answer.

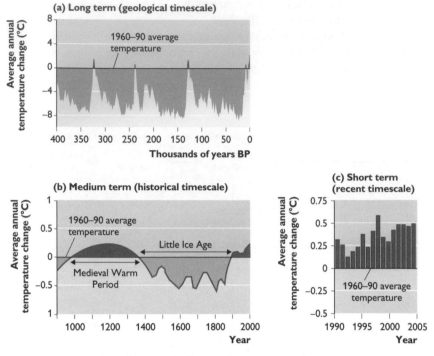

Figure 14 Climate change timescales

Climate change timescales and how the evidence is gathered

- **Long-term (geological)** — this essentially refers to the cycle of ice ages, which began some 450,000 years ago. Ice ages and the warmer periods in between them, known as interglacials, occurred at almost regular intervals throughout the geological period known as the Quaternary period.

The best evidence for long-term climate change comes from ice cores from the Greenland and Antarctic ice sheets. Air can be sampled from bubbles trapped in the ice as it was laid down over thousands of years, and carbon dioxide levels in the former atmosphere can be reconstructed. During the ice ages carbon dioxide levels are very low (180 ppm) but in the warm interglacials they rise to 280 ppm. (Present-day levels are around 384 ppm.) Further evidence is provided by pollen extracted from sediment cores from peat bogs. Pollen grains are preserved in waterlogged peat. Different-shaped pollen grains signify different trees, some of which lived in Arctic conditions (birch) and others in warmer conditions. Unfortunately, the evidence is incomplete and long sequences are rare.

- **Medium-term (historical)** For the period between AD 1000 and 1850, there is no direct climate evidence so scientists rely on a series of proxy records to reconstruct the climate. For example, evidence from paintings, poems, diaries etc. suggests that there was a 'Little Ice Age' from around 1500 to 1800 during which the Thames froze over regularly.

The pattern of retreating glaciers can also be seen from paintings of the Alps, suggesting that their most recent maximum extent was around 1850. Records of dates of French grape harvests have been used to identify warm and cold periods, and the thickness of tree rings is also an excellent indicator. In higher temperatures, with more sunlight and abundant precipitation, trees grow well and have thicker tree rings, whereas in colder periods their growth is stunted. For some trees, such as the ancient bristlecone pines in western USA, evidence goes back 4,500 years.

- **Short-term (recent)** is much more straightforward to analyse. Quality instrumental records have existed for the last 100 years, as have detailed records of the response of ice sheets and glaciers. The ice sheets of Greenland, the Canadian Arctic and Antarctica are all aerially surveyed and monitored continuously. Air and ocean temperatures are recorded and ecosystem changes are monitored because their disappearance would have such an impact on world systems.

The drivers of climate change

Natural causes are the main drivers of long-term and medium-term climate change.

It is generally accepted that **Milankovitch cycles** (orbital eccentricity, axial precision and axial tilt) are the basic cause of **long-term climate change**. These are variations in the Earth's orbit around the sun that cause a change in the amount, distribution and seasonal timing of solar radiation. The theory is named **astronomical forcing**. In support of the theory is the regular pattern of ice ages and interglacials. The actual impact of these orbital changes is enough to change global temperatures by 0.5°C, whereas records show that ice ages were about 5°C colder than the interglacials. The cycles may have been just enough to trigger these more substantial climate changes. For example, feedback mechanisms (such as greater reflection of solar radiation from increased snow cover that has a higher albedo) could tip the Earth into an ice age.

Medium-term climate change in historical times, which includes the Medieval Warm Period and the Little Ice Age, has been linked to changes in **solar output** at timescales longer than the regular 11-year sunspot cycle. Sunspots (dark spots on the sun's surface caused by intense magnetic storms) have been well recorded over the last 400 years. It would seem that the Little Ice Age could be linked to a long period with almost no sunspot activity.

Volcanic activity at a super-volcano scale can also alter global climate. Volcanoes eject huge volumes of ash, water vapour, sulphur dioxide and carbon dioxide into the atmosphere, which reduces the amount of sunlight received at the Earth's surface. However, even for a very large eruption such as that of Krakatau in the nineteenth century, the temperature changes are short-lived (1–2 years). Patchy historic records do suggest increased volcanic activity during the Little Ice Age but this is unlikely to be its cause.

Short-term climate change could be part of the longer-term cycle of natural causes. However, scientific opinion increasingly supports the view that the unprecedented scale and rate of change is linked to the huge rise in **greenhouse gas emissions** and is therefore a direct result of the enhanced greenhouse effect (**atmospheric forcing**). Concentrations of carbon dioxide, ozone, methane, nitrous oxide and chlorofluorocarbons in the atmosphere have grown significantly since pre-industrial times as a result of human activity. These contribute to the enhanced greenhouse effect (Figure 15).

Knowledge check 25

If there are many sunspots, what impact does this have on medium-term climate change?

Knowledge check 26

Why has volcanic activity been discounted as a major cause of climate change?

Knowledge check 27

Explain the difference between astronomic and atmospheric forcing.

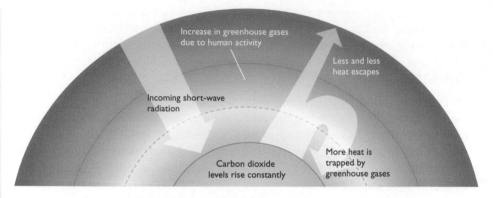

Figure 15 The enhanced greenhouse effect

Knowledge check 28

What is the greenhouse effect and what has caused it to be recently enhanced?

Eight key questions on climate change for you to consider

1 Is short-term climate change part of the longer-term cycle of climate change or is it happening at an unprecedented rate and therefore something special?

2 Is this short-term climate change caused by natural or human (anthropogenic) causes?

3 What is causing global warming?

4 Is there any interconnection between global warming and oscillations such as the El Niño Southern Oscillation (ENSO)?

5 Why are the impacts of climate change so difficult to predict?

6 How and why do they differ from place to place and how are they likely to change over time?

7 What makes climate change the world's most difficult problem to solve? Why does this cause conflict?

8 Will we reach the tipping point or come back from the brink?

The impacts of global warming

Direct impacts

The two compulsory case studies emphasise the scale of the environmental, social and economic impacts of global warming. The **Arctic** has been selected because the environmental impacts are happening here and now. It is both a 'barometer' of environmental impacts and an early warning as to what the rest of the world might face. In contrast the continent of **Africa** is only expected to have average rises of 2–3°C by 2100, but the social and economic impacts are made far worse by the sensitivity of the environment, the vulnerability of the people and their low adaptive capacity to issues such as drought, water shortages, food insecurity, worsening health, debt and endemic poverty.

Compulsory case study **Environmental and ecological impacts of climate change on the Arctic region**

In the past few decades, average Arctic temperatures have risen at twice the rate of the rest of the world (3–4°C in the last 50 years in Alaska and northwest Canada). Over the next 100 years they could rise a further 3–5°C over land and up to 7°C over the oceans. This is already leading to the melting of the Greenland ice sheet (see Figure 16). The Arctic Impact Assessment website (**www.acia.uaf.edu**) provides detailed information.

Figure 16 Arctic region: summary of key changes

Impacts on natural systems

- **Vegetation shifts.** Vegetation zones are predicted to shift northwards. This shift will destabilise existing food webs. The longer, warmer growing season will benefit Arctic agriculture although soils will be a limiting factor.
- **Thawing permafrost.** Up to 40% of total permafrost is expected to thaw, especially in Siberia. In some areas, lakes and rivers will drain. This will have an impact on species, particularly freshwater fish such as the Arctic char and lake trout.
- **Increasing fires and insects.** Global warming will increase forest fires and insect-caused tree death, which may have an impact on old-growth forest, a valuable habitat.
- **Ultraviolet impacts.** Increased ultraviolet (UV) radiation will reach the Earth's surface as snow and ice cover is lost. UV radiation destroys phytoplankton at the base of the marine food chain.
- **Carbon cycle changes.** The replacement of Arctic vegetation with more forests will lead to higher primary productivity and increased carbon dioxide uptake, but increased methane emissions could counterbalance this positive impact.
- **Other impacts.** Increased coastal erosion as thawing permafrost weakens the coast, and there are more waves and storm surges as the protection of sea ice is lost.

Knowledge check 29

Why will vegetation zones such as the Taiga (boreal forest) shift further north?

Examiner tip

Ring the key ecological words (i.e. to do with plants and animals) and the key environmental words. This is what the Arctic case study is about.

Impacts on animal species

- Species will shift north with forests. Some species are likely to suffer major decline.
- Marine species dependent on sea ice, including polar bears and ice-living seals, will decline. Birds like geese will have different migration patterns.
- Land species adapted to the Arctic climate, including Arctic fox and caribou, are at risk.

Impacts on society

The ecological and environmental changes described above will mean:
- loss of hunting culture and decline of food security for indigenous peoples
- need for herd animals to change their migration routes
- enhanced agriculture and forestry
- the Arctic will become more accessible, and vulnerable to exploitation for oil, gas, fish and other resources

Compulsory case study **Socioeconomic impacts of climate change on the continent of Africa**

Africa is the continent that makes the least contribution to global warming, yet it is the most vulnerable to climate change. Much of its population is dependent on climate-sensitive resources such as local water and ecosystems, and has a limited ability to respond to changing climate because of poverty.

It is predicted that temperatures in Africa overall will rise by around 3°C by 2100. Rainfall is likely to increase in the equatorial region but decrease to the north and south of that band.

Figure 17 summarises Africa's vulnerability to global warming.

Knowledge check 30

Explain why climate change will have such a major impact on the continent of Africa.

Examiner tip

Examine Figure 17 in detail so that for each impact you are able to locate examples, e.g. of where there will be food insecurity issues. Remember Africa is a continent with many different countries, all experiencing different impacts.

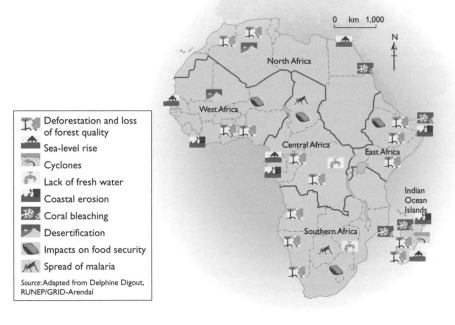

Figure 17 Global warming impacts in Africa

- **Water issues.** Life in Africa is regulated by access to water for agriculture, domestic use and hydroelectric power. Demand outstrips supply of water for 25% of Africans. However, enough water is available in most parts of Africa. Poverty is the key reason why millions have no access to safe and reliable water supplies. Water stress could lead to wars, global migration and famine.

- **Food insecurity.** Seventy per cent of the population are subsistence farmers, many of whom will not be able to feed themselves should water supplies dry up, pasture quality deteriorate or crops fails. Increased locust plagues may also threaten food supplies.

- **Natural resources.** Poor people, especially those living in marginal environments, depend directly on wild plants and animals to support their way of life. Loss of biodiversity due to climate change will threaten them.

- **Health.** Vector-borne diseases (e.g. malaria) and water-borne diseases (e.g. diarrhoea) could increase with climate change, overloading inadequate healthcare systems.

- **Development of coastal zones.** Movement of environmental refugees from the countryside puts pressure on the coastal zones; 60% of Africans live in coastal zones, many of which are at risk of coastal erosion and flooding. The threat from these is likely to increase as a result of rising sea level. If the coastal zones were flooded, much of the continent's infrastructure of roads, bridges and buildings would also be lost.

- **Poverty.** At the root of Africa's vulnerability is poverty. The problem is made worse by conflicts (e.g. Darfur). An unjust trading system forces many countries to sell their exports at a low price. Above all, the burden of unpayable debt means that no money is available for the mitigation of climate impacts and the introduction of adaptive strategies.

Indirect impact: rising sea levels

Predicting sea-level rise is complex and uncertain. It is caused by both thermal expansion of the oceans (increase in volume with warming) and melting ice sheets. Some areas are particularly vulnerable to rising sea levels:

- the world's largest deltas, e.g. Brahmaputra/Ganges in Bangladesh, Nile and Mississippi; these have large populations and high risk of exposure to storm surges, flooding and sea-level rise
- small low-lying islands (coral atolls in the Indian and Pacific Oceans)
- areas close to sea level that are already heavily defended, e.g. the Netherlands and eastern England (the Wash)
- places such as Hong Kong and Singapore, where new developments have been built on reclaimed coastal land

The first environmental refugees have already been registered from the Char Islands of Bangladesh and Tegua in the South Seas (2006).

Predicting emissions and their likely impact

The Intergovernmental Panel on Climate Change (IPCC), first formed in 1988, collects data on greenhouse gas concentrations, sea levels and ice stores. As the prediction of climate change is so uncertain, the IPCC envisages different possibilities based on a range of emissions scenarios (Figure 18).

Knowledge check 31

What is likely to be the benefit of global warming for East Africa?

Knowledge check 32

Why are rising sea levels said to be an indirect impact of global warming?

Knowledge check 33

What is an environmental refugee? Why are the Char Islands and Tegua the first locations where they have been registered?

Examiner tip

Many of you will be studying 'Crowded coasts' as an option in Unit 2, so you can use some examples that you have studied for that unit. For the Unit 1 exam, you will need several brief examples of the impact of rising sea levels on different coastlines.

- **High emissions scenario:** rapid economic growth, increasing populations, reliance on fossil fuels and business as usual.
- **Medium-high emissions scenario:** more self-reliance, increasing populations and economic growth.
- **Medium-low emissions scenario:** population growth slows, clean and efficient technology, reduction in use of fossil fuels.
- **Low emissions scenario:** local solutions to sustainability, slower rate of population increase, less rapid technological change.

Knowledge check 34

Why is it necessary for the IPCC to have a number of emissions scenarios?

Why the uncertainty?

Predictions of emissions levels and their impacts are difficult, because it is hard to predict the following:

- the level and nature of economic development, particularly in countries like India and China, which will determine greenhouse gas (GHG) emissions
- what degree of international action will be taken to reduce emissions
- the inertia in the system — even if GHG emissions stabilise, climate change will continue
- the impact of positive feedback, for example as permafrost areas thaw due to global warming the powerful GHG methane will be released, increasing warming still further

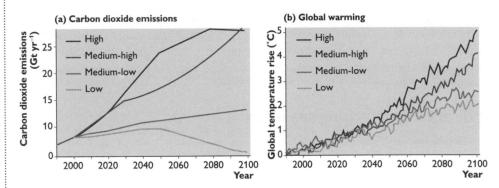

Figure 18 Climate change predictions by the IPCC

Too late to act?

The scientific evidence points to increasing risk of serious, irreversible impacts from climate change. The point when change becomes irreversible and catastrophic is called the **tipping point**.

The current level (2011) of greenhouse gases in the atmosphere is around 400 ppm carbon dioxide equivalent. Even if the annual flow of emissions did not increase beyond today's rate, the stock of greenhouse gases in the atmosphere could reach 550 ppm carbon dioxide equivalent by 2050. However, the annual flow of emissions is accelerating as fast-growing economies invest in a high-carbon infrastructure and as demand for energy and transport increases around the world. The level of 550 ppm could be reached as early as 2035. At this level there is a strong chance of a global average temperature rise exceeding 2°C, which many regard as the tipping point for temperatures. Remember: the Arctic is already showing this profile of accelerating change.

Knowledge check 35

Explain why even under a low emissions scenario, global temperatures are expected to rise by nearly 2°C by 2100.

Under a business-as-usual scenario with no serious attempt at mitigating the impacts, greenhouse gases could more than treble by 2100, giving at least a 50% risk of exceeding a 5°C global average temperature change in the following decades. This huge rise would take humans into unknown territory — disaster on an unimaginable scale.

Coping with climate change

The role of mitigation and adaptation strategies

Mitigation involves reducing the output of greenhouse gases and increasing the size of greenhouse gas sinks. It is a longer-term solution.

Examples of mitigation strategies include:
- setting targets to reduce carbon dioxide emissions
- developing energy-efficient strategies in all economic sectors
- switching to renewable energy sources such as wind power
- 'capturing' carbon emissions from power stations and storing them
- enhancing the size of carbon sinks, for example by afforestation

Arguments for mitigation are strong as it is necessary to avoid irreversible climate change and mitigation is the ultimate long-term solution. For mitigation to work, it has to take place at all scales, from local to global. Figure 19 shows the scales of mitigation and the key players.

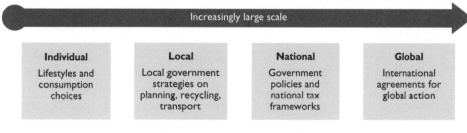

Figure 19 Scales of mitigation

Adaptation means changing our lifestyles and businesses to develop strategies to cope with the increasing dangers from climate change — rising temperatures, more extreme weather, increasing water, food and health insecurity, and rising sea levels. It is an immediate solution.

Examples of adaptation strategies include:
- coastal defences or managed retreat of a coastline vulnerable to sea-level rise
- developing a second generation of drought-resistant GM crops
- enlarging conservation zones or developing wildlife corridors to allow for shifting ecosystems and habitats

There are strong arguments for adaptation (i.e. climate proofing). Many scientists argue that the effects of climate change are happening already (as in the Arctic) and even if humans stopped polluting the atmosphere now climate change would continue. Adaptation and mitigation go hand in hand and both are needed.

Examiner tip

In the next section (on coping with climate change), a number of solutions are outlined. Beware of 'total doom and gloom'. The following websites are useful for reference: **www.iied.org** and **www.oxfam.org.uk**.

Knowledge check 36

Give one example of a mitigation strategy for each of the four scales shown.

Knowledge check 37

Explain why adaptation is seen as a more immediate solution than mitigation.

Examiner tip

For further details of mitigation and adaptation strategies, explore the IPCC website.

Key players in managing climate change

Kyoto Protocol: governments reaching global agreement?

At the 1992 Rio Earth Summit, the UN Convention Framework on Climate Change was agreed. It aimed 'to achieve stabilisation of greenhouse gas concentrations at a low enough level to prevent dangerous anthropogenic interference with the world's climate systems'. It led to the agreement of the Kyoto Protocol in 1997.

However, putting this framework into practice led to extremely complex and ongoing negotiations. Many countries delayed signing the treaty because they felt cutting emissions would damage their economies. Russia did not sign until 2005 and Australia, the second largest per capita polluter, did not sign until 2007. The USA, which produces 25% of global emissions, initially signed but withdrew in 2002 following the election of George Bush.

Complex systems were introduced allowing 'trading' of carbon credits: that is, buying unused emissions from other countries or businesses. Carbon sinks, such as planting forests, were allowed, so that countries can 'offset' emissions. Critics argue that both of these systems allow polluters to continue to pollute.

Overall reductions in emissions have been under 1%, largely because of increasing emissions from countries such as China, India and Brazil, which were not obliged to cut emissions. Discussions at the 2009 Copenhagen Climate Change Conference proved very complex and many issues surrounding mitigation responsibilities remain unresolved.

Big businesses

Some transnational corporations (TNCs) are bigger economically than developing countries and their actions wield enormous economic power. Initially, big oil and car companies were opposed to cutting emissions, arguing that reducing pollution would cost money, profits and jobs. Now, however, leading TNCs produce annual environmental reports and many are committed to greener growth. For example, General Electric, the world's largest company, has committed itself to double its research and development spending on clean technologies, double its revenue from renewables and energy-efficient products and reduce its greenhouse gas emissions by 30% by 2012. This change of attitude can be explained by pressures from customers and investors and the need to display a good public image. There are big profits to be made in renewables and new clean technology. Incentives are available for offsetting schemes and clean development mechanisms, and governments are increasingly taxing carbon emissions.

National governments

National governments play a vital role in developing strategies to reduce emissions, by:
- promoting energy-efficient technology, e.g. ecobuildings, green transport
- encouraging energy generation from nuclear power (France), renewables (Scandinavia and UK) or biofuels (USA) instead of fossil fuels
- promoting carbon storage by sequestration of power-station emissions (Norway) and creation of sinks by tree planting

Examiner tip

Research the Copenhagen and Cancun Climate Conferences so you can update your knowledge and see whether progress has been made or not.

Knowledge check 38

What is meant by greener growth?

- awarding huge grants for developing carbon-neutral ecotowns
- legislation and fiscal policies, e.g. taxing pollution

Local governments

Local governments tend to reflect the strategies of their national governments. In the UK, London has led the way with legislation such as the congestion charge. Leicester is the UK's first environmental city. Figure 20 shows a range of emission-reducing schemes available to local authorities.

Figure 20 How a city can reduce greenhouse gas emissions

Areas throughout the world have developed Local Agenda 21, which emerged from the 1992 Rio Earth Summit. It encourages local authorities to implement a raft of sustainable strategies to improve the environment and reduce carbon emissions. Many people are encouraged by positive local actions on recycling, green transport and small-scale schemes (e.g. combined heat and power) to do their bit for climate change.

NGOs

Most environmental NGOs operate in advocacy and education roles to bring pressures on governments and individuals to act. They campaign on the causes, impacts and solutions to climate change for a number of environmental and socio-political reasons.

Individuals

The key to getting individuals involved in cutting emissions is education. Calculating the **carbon footprint** (the amount of carbon dioxide emitted by an individual or a family or school each year) is a very good personal check on performance. It creates an opportunity to reduce the footprint by changing patterns of consumption and transport use, by installing energy-efficient systems and, if all else fails, by offsetting carbon-emitting actions. However, many of these changes may be challenging, sometimes even impossible to implement. Think about some of the following ideas:

- walking, cycling or taking the bus/train to work or college, i.e. not using the car
- buying locally produced food — often at higher prices and with limited choice
- switching energy supplies to renewables — this is costly
- becoming energy efficient in the home, e.g. not leaving the television on standby, not using the tumble dryer in summer

Knowledge check 39

Chose four of the strategies shown in Figure 20 and explain how they help cities reduce GHG emissions.

Examiner tip

Research the Sustainable Cities website and add to the strategies.

Examiner tip

Look at the websites of some NGOs which campaign on climate change (e.g. Camp for Climate Action, Eco Equity). Find out what their role is and how they campaign. Assess the degree of bias in their information.

Knowledge check 40

How does a carbon footprint differ from an ecological footprint?

The challenge of global hazards for the future

The enormity of the challenge

Global warming and oscillations like El Niño and La Niña are causing rising numbers of hydro-meteorological hazards (wind storms, drought and floods). The increasing numbers of vulnerable people, who have low adaptive capacity, mean that these hazards frequently develop into disasters.

Water shortages

Climate change will exacerbate water stress (lack of a safe, reliable supply), which is currently experienced by some 40% of the world's people. As a result of rising temperatures, greater incidence of drought or loss of supplies from glaciers, around half a billion more people will experience a physical shortage of water, i.e. demand exceeding supply. As poverty levels rise, there may be many more people not able to afford clean water, i.e. an economic shortage of water. Water is needed not only by people themselves but also for their rainfed agriculture and stock. Water shortage will have a profound effect on food security.

Food insecurity

Food insecurity means people's inability to access an adequate diet. Climate change will exacerbate famine in parts of Africa, South America and south Asia, as an increase in extreme weather events leads to falling yields in cereals. Rising poverty will mean that fewer people can afford to access food and more will rely on food aid. Climate change has not only led to reductions in crop yields. The desire to cut emissions has also led to the planting of millions of hectares of maize/soy beans/sugar cane for **biofuels** at the expense of growing cereals for food.

The relationship between global warming and food security is complex. The growing of cash crops (notably tea) in some regions will be affected by changing conditions, but elsewhere there will be some benefits, such as a longer growing season in the Arctic and more rainfall in Kenya.

In theory, agriculture should be able to adapt to climate change using a variety of technologies, from intermediate technology such as soil conservation schemes in Mali, to high technology such as the long-awaited second generation of drought-resistant/salt-tolerant GM crops. The reality is that ecosystems are extremely sensitive, the vulnerability of people high, and the adaptive capacity limited in the developing world.

Tackling the challenge

A number of sustainable strategies can be used to tackle climate change. However, many of them have costs as well as benefits. Green strategies include a range of so-called 'eco-friendly' projects.

Tree planting

Trees act as a 'sink' by taking in carbon dioxide and 'fixing' it in the form of hydrocarbons (plant matter). However, the benefits are not felt for at least 10 years

Knowledge check 41
What is meant by rainfed agriculture?

Knowledge check 42
What is the difference between physical and economic water shortage?

Knowledge check 43
What are biofuels?

Knowledge check 44
State one reason why biofuels are considered eco friendly and one reason why their growth is causing such concern.

Knowledge check 45
What is intermediate technology?

Knowledge check 46
Why are second generation GM crops long awaited?

as a growing tree releases more carbon dioxide than it absorbs, especially if the ground vegetation is burned for the initial tree planting. Under the Kyoto Protocol, countries are not prohibited from deforestation, yet they can claim carbon credits for new planting, while in reality releasing carbon dioxide.

Renewable energy projects

These cut down on the use of fossil fuels. Some small-scale projects — such as solar panels, microhydro turbines, biogas converters and wind turbines — are of vital importance to remote areas in developing countries, revolutionising the quality of life of the rural poor.

Any large-scale schemes, such as the Three Gorges dam in China or the new mega dam proposed for the River Congo, have all kinds of environmental and ecological costs, as well as the obvious benefits of providing large quantities of 'green' electricity along with flood control, irrigation and tourism. In some countries that depend on hydroelectric power (HEP), there is concern that climate change will render HEP schemes useless because of drought affecting water supplies.

The most contentious development is probably that of **biofuels**. There are two groups of the so-called first generation of biofuels: **biodiesel** is usually made from soya beans or rapeseed, which is added to conventional diesel; **bioethanol** is made from corn (USA) and sugar cane (Brazil). There are huge concerns about the environmental damage and impacts on biodiversity resulting from destruction of rainforests or draining of wetland areas to grow these biofuel crops in countries such as Brazil or Indonesia. They also exacerbate the food-security crisis where they are grown instead of food crops such as wheat or rice. The question has to be asked whether the costs of growing biofuel crops outweigh the benefits and whether it is a truly 'green' fuel. In the USA, the main driver seems to be combating the rising price of oil rather than reducing greenhouse gas emissions.

Wind farms produce green electricity but large areas of turbines are required to generate comparatively modest amounts of power. Some of the best sites in terms of wind occur in areas of outstanding natural beauty and development of wind farms is hotly opposed in such locations. Recently a large scheme in the peatlands of the Isle of Lewis in the Outer Hebrides was rejected on environmental grounds, with 11,000 residents signing a petition against it.

A major problem with renewable sources of energy is that they tend to be more expensive than fossil fuels. Nuclear power is another option, but opinion is sharply divided as to whether it is a *green* option, because of waste and safety issues.

Community-based strategies

These often work well as they are developed in a 'bottom-up' fashion by the community, as opposed to being forced upon them by higher authorities — that is, they are proactive. Communities in countries at all stages of development can take the initiative in producing schemes that mitigate or adapt to climate change. The only problem about bottom-up schemes is that they are often locally based and not easily replicated everywhere. Their success also depends on the energy of a 'hard core' of leaders and organisers to motivate the community.

> **Examiner tip**
> Use the Practical Action website to read up about some of these rural small-scale renewable energy projects.

> **Examiner tip**
> Always look at the pros and cons of all renewable energy schemes.

> **Knowledge check 47**
> Why do people support bottom-up strategies?

Examples include the creation of low carbon communities in towns and villages such as Wolvercote (Oxfordshire), and polythene-bag-free towns such as Modbury (Devon). There are numerous community-based initiatives, frequently financed by NGOs, in developing countries. These are known as **capacity building schemes**, which teach local people to adapt to climate change. Community forest projects are also widespread.

Energy efficiency

This is possibly the most effective way forward, although initial costs of any scheme are high. Long-term energy efficiency not only reduces emissions, but it actually cuts costs and improves local pollution levels. Avoiding dirty development in the first place is clearly the way ahead for developing countries.

Methods of increasing energy efficiency include:

- remodelled factories with clean, efficient industrial processes and optimum use of energy
- redesigned houses with modern boiler systems and full insulation
- greener and more efficient power stations
- green transport using new, greener fuels (hybrid technology or hydrogen-powered)
- use of recycling, e.g. waste heat from incineration for central heating systems

One word of warning: India and China, because of their huge size (their combined populations comprise one-third of the world's people, 2.5 billion) and rapid economic growth (average of 6–10% per year) have a critical role to play in controlling emissions. Even if all the more developed countries cut their emissions drastically the following statistics suggest that the economic growth of China and India will lead to more emissions:

- The increase in China's greenhouse gas emissions from 2000 to 2030 will nearly equal the increase from the entire industrialised world. In 2008 China became the world's number one greenhouse gas emitter.
- India's greenhouse gas emissions are estimated to rise by 70% by 2025, with India's energy consumption rising even faster than China's.
- Both India and China are desperate for energy to fuel their economic expansion, in order to bring their citizens out of poverty and improve the quality of their lives — 40% of India's population currently lack regular access to electricity. Environmental concerns inevitably lag behind the need for growth, but both countries are concerned about pollution and energy shortages.

Solutions in a more hazardous world

The Hyogo Framework for Action (Figure 21) was developed in 2005 by the World Conference on Disaster Reduction held in Kobe, Japan. At its heart is risk reduction by building resilience to disasters and by overcoming the underlying factors that lead to vulnerability, notably poverty. Their aim is to exploit the Kyoto mechanisms, such as the Special Climate Change Fund and the Clean Development Mechanism, to allow developed countries to pay for greenhouse gas cutting projects in return for carbon credits.

Knowledge check 48

Why is energy efficiency the most effective way forward?

Examiner tip

China has begun to make huge progress as a developer of wind and solar technology. Find out about China's strategies for the future and the impact these could make on GHG emissions.

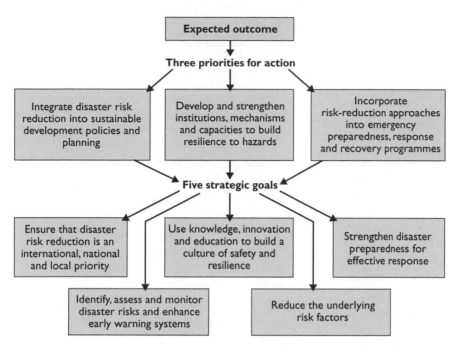

Figure 21 The Hyogo Framework for Action

Knowledge check 49

How did the Hyogo Framework get its name?

The framework can be applied at all scales, from local to international. In Unit 2 you will be investigating recurring hazard events — either river flooding or the impact of coastal storm surges. At an international scale, a mega-disaster such as the 2004 south Asia tsunami can be researched, looking at how the new coastal zoning and coastal economic development schemes, and the Indian-built tsunami early warning systems, have fulfilled the three priorities for action in the Hyogo Framework. At a regional scale you could research poverty in sub-Saharan Africa to see how the various countries are using the framework and at a local scale you could look at flooding and the management of flood risk (using the Environment Agency website: **www.environment-agency.gov.uk**).

In conclusion, Figure 22 summaries the importance of an integrated approach to management of the rising risk from climate change and hydro-meteorological hazards.

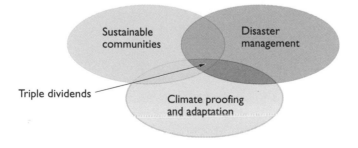

Figure 22 The importance of an integrated response to risk management

Knowledge check 50

Explain how the triple dividends result from the response shown.

Summary

World at risk

- Rising risks result from an increasingly unfavourable risk equation. The increasing impacts of short-term climate change have led to extreme weather causing more hydro-meteorological hazards. The level of vulnerability is increasing, whereas people's capacity to cope is decreasing, especially in developing nations.

- **Global trends:** the number of disasters is increasing, especially hydro-meteorological disasters. The number of people affected has increased. Deaths from disasters have levelled out, largely because of improved management of major hazard events. Economic losses have dramatically increased.

- **Global patterns:** the distribution of geophysical hazards is largely explained by plate tectonics. The distribution of hydro-meteorological hazards is explained by the occurrence of particular meteorological and climatic conditions. Slides can be caused by geophysical activity and extreme weather events in mountain areas. Where several different types of hazard occur in an area with vulnerable people, multiple disaster hotspots occur. An in-depth study of California and the Philippines illustrates this.

- **Climate change** can be classified into: long-term change caused by the Milankovitch cycles — evidence comes from ice cores. Medium-term historical change is thought to be caused by sunspot activity — evidence comes from proxy records. Recent short-term climate change is caused by oscillations such as El Niño/La Niña, and global warming. The latter is increasingly linked to anthropogenic causes, i.e. rising levels of GHGs. Evidence comes from direct measurements.

- The Arctic is experiencing ecological and environmental impacts as temperatures rise at twice the rate of the rest of the world.

- The continent of Africa shows how the vulnerability of the people can magnify the socioeconomic impacts.

- Indirect impacts include rising sea levels.

- Climate change is an uncertain science as predicting the level of GHG emissions and their likely impact is complex.

- Solutions to climate change include shorter term local adaptations and longer term mitigation.

- Solutions rely on a range of actions from key players operating at all scales from local to global.

- Chronic hazards such as short-term climate change provide a huge challenge for the future and rely on the development of sustainable strategies for disaster management.

Going global

'Going global' examines the process of globalisation: what drives it and its winners and losers. Increasingly rapid social and economic change has accompanied globalisation, altering family structures, work patterns and jobs. Increasing mobility has encouraged migration on an unprecedented scale, especially to cities. 'Going global' explores the benefits and costs of an increasingly global world and examines some of the challenges and potential solutions. Figure 23 shows how the topic content links together.

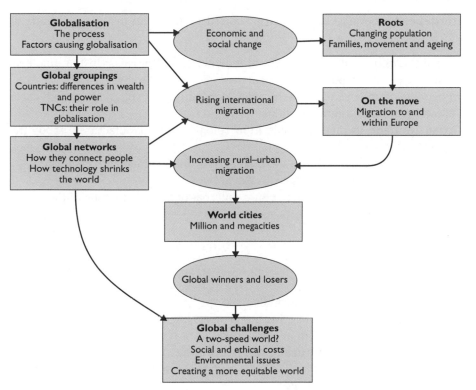

Figure 23 Topic concept map

A key theme of globalisation is the quickening pace of technological change. Figure 24 contrasts the gradual adoption of older technologies with the extraordinarily rapid growth of mobile phones and the internet. The first UK text message was sent in 1992 and broadband only became widely available in 2000.

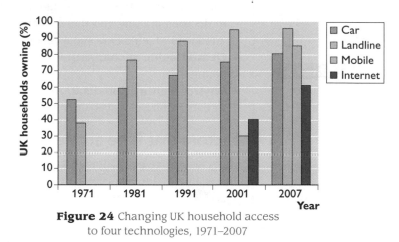

Figure 24 Changing UK household access to four technologies, 1971–2007

Life without broadband, smart phones, iPods and social networking sites might be hard to imagine, but only 10–15 years ago none of these existed.

Globalisation

What is globalisation?

The International Monetary Fund (IMF) defines globalisation as 'the growing economic interdependence of countries worldwide through increasing volume and variety of cross-border transactions in goods and services, freer international capital flows, and more rapid and widespread diffusion of technology'.

Interdependence suggests that in a globalised world, events in any one country will quickly spread to others.

Economists argue that globalisation is a centuries-old process. In the nineteenth century the British Empire was a 'global network' connected by:

- a network of undersea telegraph cables, connecting to overland telegraph cables, allowing for rapid communications
- extensive trade in raw materials from colonies and manufactured goods from the UK
- a shared head of state (Queen Victoria for most of the nineteenth century)
- global military power, especially the Royal Navy
- attempts at imposing a shared culture (international cricket is one of the clearest links back to imperial times)

Types of globalisation

Globalisation takes several forms, and all involve increasing interdependence and interconnectedness:

- **Economic globalisation**, including the growth of transnational corporations (TNCs), with a global presence and brand image, the global spread of foreign direct investment (FDI) and growth in world trade.
- As TNCs have grown, **cultural globalisation** has followed. Increasingly people eat similar fast-food, such as Subway and McDonald's, and watch global news channels such as Sky and CNN.
- **Political globalisation** is the dominance of Western democracies (G8 countries) in decision making and the view that democratic, consumerist societies are the 'model' to which others should aspire.
- **Demographic globalisation** results from increasing migration and mixing of populations.
- **Environmental globalisation** involves the realisation that global environmental threats require global solutions.

Factors that have accelerated globalisation

Many factors have contributed to our increasingly globalised world. Table 3 explains and evaluates the importance of these factors.

Examiner tip

You should learn definitions of key terms such as globalisation as part of your exam preparation. Keep a glossary of new terms in your file to use during revision. Some exam questions are straightforward definition questions.

Knowledge check 52

What is foreign direct investment (FDI)?

Table 3 Factors encouraging globalisation

Factor	Explanation
Free trade	Removing trade tariffs and quotas has promoted easier and faster trade.
International organisations	The World Trade Organization has promoted free trade, and trade blocs have encouraged free trade within and between groups of countries.
TNCs	TNCs have shifted production to the developing world and created global connections and trade links.
Communications technology	Satellite and fibre optic communications have led to growth of mobile phone and internet communications, and falling communication costs.
Transport technology	Containerisation since the 1960s has led to increasingly cheap, automated and efficient methods of transport. Ships can carry 9,000+ containers. For high-value, low-volume goods and people, cheap air travel has revolutionised transport.
Financial deregulation	Government controls on banks, currencies, interest rates and companies have all decreased. This makes investment easier and profits higher.
Consumers	The global consumer has contributed to soaring demand for goods from all corners of the world.
The media	Large global media corporations have global reach and present a similar 'world view' of the news, contributing to the sense of a connected world.

Knowledge check 53

How has transport technology contributed to globalisation?

Opportunities

Globalisation presents countries, companies and individuals with numerous opportunities. Great wealth has been created through trade. China's exports rose from $200 billion in 2000 to over $1,200 billion in 2009. Foreign direct investment has led to industrialisation, resource exploitation and property development. This creates jobs and wealth, plus a return for the investor. FDI into China increased from $4 billion in 1991 to $90 billion in 2009. Individuals have benefited through migration to economic hotspots such as Dubai and Bangalore. Developed world economies are increasingly knowledge driven — information is exchanged, rather than goods. Investment in education is one way of increasing opportunities, another is 'poaching' key people, increasingly from all corners of the globe.

While many people welcome globalisation, it does have drawbacks:
- It has led to growing inequality — the rich have grown richer, and the poorest relatively poorer.
- It is a process dominated by TNCs and governments, and ordinary people may feel like 'pawns' (e.g. lack of unions, loss of jobs due to global shifts and exploitation of workers in developing nations).
- It encourages unsustainable economic growth, excessive resource and energy use, and therefore has negative environmental consequences.
- It tends towards a 'Westernised' global culture where local traditions, languages, food and art disappear.

Population movements

Globalisation causes migration. Most migrants move for economic opportunity. The movements may be legal or illegal. As some parts of the world have become wealthier, the gap between the wealthy and the poorest has widened. This encourages migration for greater opportunities. The world migrant stock in 2008 stood at about 214 million:

- 145 million legal economic migrants (80 million in 1970)
- 20–30 million illegal migrants (largely economic)
- 15 million refugees
- 26 million internally displaced people

When migrants cannot move legally, they move illegally. These movements are towards economic hotspots and are usually from the developing to the developed world:

- the USA — major flows from Mexico, including up to 10 million illegal immigrants across the US–Mexico land border
- the EU — including up to 5 million illegal migrants across the southern EU fringe of Spain, Malta and Italy
- the Middle East — attracts large numbers of migrants from south Asia to work in construction and domestic service

Global groupings

Rich, developed countries have retained their dominance during the era of globalisation. A lucky few countries have joined this rich group, for example the 'Asian Tigers' of Singapore and South Korea. Economic power is concentrated in the three global economic cores of North America, Europe and Asia. Of the 500 largest TNCs in the world in 2010, 139 are based in the USA and 71 in Japan. The three cores are referred to as the **triad**. Around 80% of all global economic wealth and trade is concentrated in this triad, which is linked through a complex system of global finance, stock exchanges, international airports and government centres.

Economic groupings

Geographers classify countries by their level of economic development, specifically their **GNI** (gross national income). Dividing the world into the 'rich North' and 'poor South' is a common device — but the world is more complex than this. Figure 25 shows that between 1975 and 2002 some groupings of countries became wealthier, but others stagnated or became poorer. This has created a world of the rich, those getting richer, and those either stagnating or getting poorer.

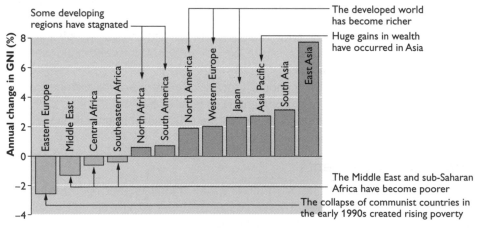

Figure 25 Changes in GNI, 1975–2002

Economic groupings are useful in geography because they allow us to compare different development pathways. Countries can, and do, move between groups. Singapore was once classed as an NIC but is now a fully developed country. Table 4 outlines the most commonly used economic groupings.

Table 4 Economic groupings of countries

Economic grouping	Examples	World Bank groupings by annual per capita income, 2006	Economic structure
More economically developed countries	UK, USA, Australia, Greece, Singapore	High income (more than $11,115)	70%+ of jobs are tertiary. TNCs are based here. R&D and quaternary industry important.
Newly industrialised countries	South Korea, Brazil, Taiwan, Chile, China	Upper middle income ($3,596–11,115)	30%+ of jobs are secondary and manufacturing exports important.
Recently industrialised countries	Thailand, Indonesia, Tunisia	Lower middle income ($906–3,595)	High primary employment. In key cities, manufacturing is growing rapidly.
Less economically developed countries	Egypt, Peru	Low income (under $906)	Primary industry provides 40%+ of jobs. Development is slow.
Least developed countries	Malawi, Bangladesh, Haiti		Dependent on primary industry. High levels of poverty persist. Many LDCs getting poorer in absolute terms.

Political groupings

Formal groupings of countries, called inter-governmental organisations (IGOs), exist where countries have signed agreements for economic gain. Such groups contain 'like-minded' countries usually at similar levels of development. In most cases the groups protect the interests of member states (Table 5).

Knowledge check 56

Which global political groupings does the UK belong to?

Table 5 Political groupings of countries

Grouping	Members	Role	Global importance
European Union (EU)	Formed in 1957 27 European countries, including the UK	Economic union, with internal free trade and migration	7.5% of world population but 31% of total global GDP
Organisation for Economic Cooperation and Development (OECD)	Formed in 1961 30 members, 25 of which are MEDCs. All are democratic market economies	The OECD monitors economic performance and works to reduce corruption and bribery	75% of global GDP is generated within the 30 member countries
Organization of the Petroleum Exporting Countries (OPEC)	Formed in 1960 12 major oil exporters in the Middle East (e.g. Saudi Arabia, Kuwait), South America and Africa	OPEC safeguards the interests of oil exporting countries. It has a large influence on the global oil price	OPEC countries have 65% of global oil reserves and 35% of production
Group of Eight (G8)	Formed in 1974 UK, USA, France, Canada, Germany, Italy, Japan, Russia	An informal forum of the super-rich and powerful countries	15% of world population but 65% of global GDP
Group of 20 (G20)	(G20) Formed in 1999 19 countries, including China, India and Brazil, plus the EU	Increasingly relevant following the global recession in 2008–09; many G20 NICs are growing rapidly and are cash rich	Accounts for 85% of global GDP
Group of 77 (G77)	Formed in 1964 Most African, Asian and Latin American nations (131 in total)	A grouping of developing nations formed to give a collective voice to the developing world	Its influence is lessening as the G20's influence has grown

Trade blocs

Political groupings usually have economic motives, to press for the interests of their member states. More formal trade blocs, such as the European Union (EU), North American Free Trade Area (NAFTA) and Association of South East Asian Nations (ASEAN), allow free trade, without taxes, tariffs or quotas between member states. Countries wishing to export into the bloc often have to pay trade taxes and/or have quotas imposed. These external trade barriers protect trade within the bloc.

The World Trade Organization (WTO) works to reduce trade barriers and create free trade between blocs. Increasingly the world is caught between these two forces. Trade blocs have tended to become more common, while free-trade agreements have also increased. There is general consensus that free trade is good, but increasingly an unwillingness to make it more free. The current set of talks to increase free trade, the Doha Round, began in 2001, but agreement has not yet been reached.

International trade growth has led to some significant shifts in wealth and power:
- Developed nations have maintained their 'top slot'.
- The 'Asian Tiger' NICs have developed almost to developed-country levels.
- In the last 10 years, Brazil, Russia, India and China (the 'BRICs') have all gained economic power and wealth.
- Asian and Latin American NICs and RICs have grown but often in a 'boom and bust' fashion.
- Many African countries have barely benefited, with population growth outstripping economic growth, leading to income stagnation.

Knowledge check 57

Which global region/continent has benefited the least from the growth in global free trade over the last few decades?

The role of TNCs

Transnational corporations (TNCs) are major companies with a global 'reach' and a presence (production, HQ, sales) in at least two countries. TNCs are economically powerful and politically influential, and are important creators of wealth. As Table 6 shows, the largest TNCs have turnovers (sales) equivalent in size to large countries. Some, such as Walmart, employ enough people to populate an entire city. This gives them huge power. Investment decisions made by TNCs can be both a blessing and a curse.

Table 6 The world's five largest TNCs compared with countries

World's five largest TNCs, 2010	Turnover in 2010 ($ billion)	Employees 2009/10	Equivalent country (total GDP $bn 2010)
Walmart Stores	408	2,100,000	Norway (413)
Royal Dutch Shell	285	101,000	Venezuela (285)
Exxon Mobil	284	83,600	Columbia (283)
BP	246	80,300	UAE (239)
Toyota Group	204	320,000	Nigeria (206)

Knowledge check 58

Which industrial sectors are the top five TNCs in Table 6 involved in?

Growing TNCs

The US retailer Walmart opened its first store in 1962 in Arkansas. By 1980 it had expanded to around 300 stores in 11 states. In 1991 Walmart opened in Mexico City, and became international. TNCs focus on expanding their global operations in order to grow. This can be achieved in a number of ways.

- Acquisitions: Walmart expanded by taking over Asda in the UK in 1999, and Woolco in Canada in 1994.
- Joint ventures: in 2007 Walmart and Indian company Bharti Enterprises announced plans to jointly open Bharti Walmart stores in India.
- Franchising: 75% of McDonald's restaurants worldwide are franchises; local entrepreneurs pay fees to the parent company for the right to set up a McDonald's.

Table 7 McDonald's and Walmart compared

Walmart	Data for 2009/2010	McDonald's
Food and household goods retailer based in the USA	Business sector	Fast food restaurants and cafes based in the USA
2,100,000 $408 billion 8,500 stores in 15 countries	Employees Turnover Global network	1,700,000 $24 billion 32,000 outlets in 117 countries
Mega-retail stores damage smaller, locally owned stores. Consumers benefit from lower prices. Low pay and lack of union representation. Large carbon footprint due to global supply chain fuel use.	Corporate issues and criticisms	Accused in the past of promoting deforestation for cattle grazing land. Health issues, e.g. obesity, in relation to sugar and fat content of its food. Low paid 'McJobs'

Walmart and McDonald's have vast supply chains consuming huge quantities of farm produce and packaging materials. Both have made efforts to reduce resource consumption. About 30% of McDonald's packaging is now recycled. Walmart reduced its corporate carbon dioxide emissions by 5% between 2005–2008 and is aiming for 20% by 2012.

Global expansion is not as simple as exporting the same brand and products worldwide. Local laws, tastes, incomes and traditions mean that companies must adapt products for new markets. This is called glocalisation. One of the best examples of this is McDonald's:

- In the UK, McDonald's advertises all of its milk as organic and, along with the beef it uses, everything comes from British or Irish farms. Consumers value this local sourcing of ingredients.
- In Australia and New Zealand, stand alone McCafes compete with coffee shop chains.
- Menus are adapted to local tastes — halal meat and McArabia chicken wraps in the Middle East, or chicken Maharaja-Macs and vegetarian options in Hindu India.

Increasingly, TNCs have brands that are recognised worldwide. In individual countries and regions products are adapted to meet the needs of local tastes and maximise sales.

Positives of TNCs

- **Jobs:** the 200,000+ technology workers in Bangalore are employed by TNCs including IBM, Cisco Systems and Google.
- **Trade:** China's economic growth has resulted from TNCs locating manufacturing plants in its Free Trade Zones, boosting exports.
- **Connections:** TNCs' complex global networks create connections that tie local and national economies into the global economic system.

Examiner tip

Use flash cards or spider diagrams as part of revision to help you learn key facts and figures for examples and case studies. Use these facts in your answers because they make your work precise and detailed.

Knowledge check 59

Explain what is meant by glocalisation.

Negatives of TNCs

- **Exploitation:** TNCs have been accused of exploiting workers in the developing world by paying very low wages in 'sweatshop' conditions.
- **Unemployment:** Outsourcing jobs to the developing world can lead to job losses in developed countries.
- **Identity:** Local cultures and traditions can be eroded by TNC brands and Western ideas.

Costs and benefits of TNCs

Transnational corporations, with their ability to shift production from developed countries to new developing locations, have costs and benefits (Table 8). When TNCs invest in developing countries (foreign direct investment), they may also reduce production in their home country.

Table 8 Costs and benefits of TNCs

	Costs	Benefits
TNC source country (developed world)	Job losses due to outsourcing of jobs.	Environmental pollution from factories is 'exported' overseas.
	Abandoned production locations create derelict land.	TNC growth leads to higher profits, and more tax being paid.
	Outsourcing TNCs may become unpopular and suffer negative media coverage and falling sales.	Strong global companies are successful, securing HQ and research and development jobs.
TNC host country (developing world)	TNCs may pay no, or very low, taxes.	Economic growth due to job creation and rising consumption.
	New locations become polluted as environmental laws are weak.	Falling levels of poverty.
	Workers on low wages, with long hours, may be exploited and mistreated and suffer poor health.	Local supply chains may be created, leading to further job creation and business growth.
	TNCs may out-compete local suppliers, forcing them out of business.	TNCs demand infrastructure and communications that may benefit local people.

Global networks

Global connections tie the world together as never before, through:
- mobile phones, even in remote areas
- instant personal and business internet communication
- low-cost jet air travel
- cable and satellite media channels
- favourite brands — Coca-Cola, Subway, Nike — available globally

It is easy to forget that this connectivity is not available to everyone. Living in the UK with an annual income of £15,000 might allow access to most of the connections, but approximately 6 billion people worldwide live on lower incomes. This disparity means that some places benefit from global networks and become 'switched on' while others remain poor and 'switched off'.

Switched on: air travel

The first commercial jetliner, the De Havilland Comet, could carry up to 100 passengers a distance of 5,000 km in the 1950s. The 2007 Airbus A380 can carry up to 850 people a distance of 15,000 km. Low-cost airlines such as EasyJet and Ryanair have made air travel available to a wide market and are spreading to the developing world, e.g. India's Air Deccan launched in 2003.

Knowledge check 61

Suggest the negative environmental consequence of the rise in global air travel and the increase in the number of low-cost airlines.

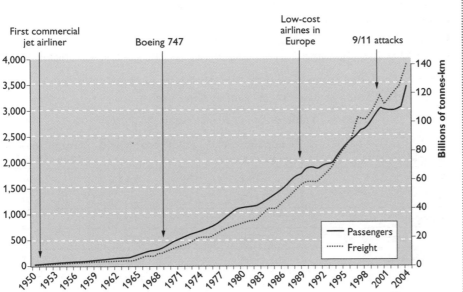

Figure 26 Global air travel growth

Increased air travel has a number of effects. It:
- reduces perceived distance between places
- increases cultural mixing, transforming exotic places into 'normal' experiences
- allows workers to move cheaply to areas of demand
- allows TNCs to move personnel and goods between production and sales locations

Air travel connects some places more than others, with the greatest connection density being between 'global hubs' in North America, Europe and Asia.

Knowledge check 62

What type of goods might be carried by air freight?

Switched on: cores and hubs

Global hubs are world cities or 'global regions'. All are in the economic core areas of the world and are centres of great wealth.

Hubs are extremely well connected to each other — for example, the financial, TNC and travel connections between London and Tokyo. Hubs act as regional nodes integrating the global economic cores such as the EU and connecting them to other cores. TNCs from hubs invest in emerging hubs, such as Bangalore in India and Shanghai in China, thus creating new connections.

Linkages between governments, TNCs and international organisations have produced a global super-rich class of media executives, lawyers, accountants, PR people, bankers and retailers who live in hubs and hop between them.

Knowledge check 63

Explain what is meant by the term 'global hub'.

Switched on: the internet

Around 1.9 billion people were using the internet by 2010. This is extraordinary for a technology that was rare in the mid-1990s. However, growth has been uneven, creating a 'digital divide' between the developed and developing world. The divide links to income levels because the internet requires:

- a PC or laptop
- a complex and reliable telecommunications infrastructure
- internet service providers (ISPs)
- the ability to pay for a connection
- companies willing to invest in providing services, such as online booking and shopping

Table 9 shows that usage growth in the developing world is rapid, but low density. Africa remains the big 'loser' in terms of internet use.

Table 9 Estimated internet use in 2010

World regions	% population (penetration)	% of total world users	Usage growth 2000–10 (%)
Africa	10.9	5.6	2,350
Asia	21.5	42	620
Europe	58.4	24.2	350
Middle East	29.8	3.2	1,800
North America	77.4	13.5	150
Latin America/Caribbean	34.5	10.4	1,000
Oceania/Australia	61.3	1.1	180
World total	29.7	100	450

Knowledge check 64

Use Table 9. Which are the most connected and least connected global regions in terms of internet use?

Examiner tip

In the exam be careful when discussing 'Africa'. While it is the most 'switched off' continent, 10%+ of people do have internet access and many African cities such as Nairobi and Lagos are increasingly connected.

The benefits of the internet are numerous and it has a significant role in development, providing:

- education and health advice
- reduced isolation of rural areas
- access to the best price for buying and selling goods
- business opportunities such as online shopping
- communication with branch plants and customers
- 'technological leapfrogging' (the bypassing of technological stages that others had to go through)
- exchange of political ideas and information (although some countries like China, Myanmar and North Korea may restrict access to this information)

Switching on through tourism

The internet may have reduced the need to travel, but it seems to have increased the demand for travel. It allows 'exploration' of distant places, plus online booking of holidays, flights and tours. Tourism business has grown faster in the developing world than the developed, as lower-cost air travel and a desire for 'something different' has led people to travel to more exotic locations. Some developing countries gain up to 40% of their GDP from tourism. Many are tourist hotspots that have seen rapid growth (e.g. Vietnam, Cuba and Madagascar).

Getting switched on through tourism requires less physical infrastructure than complex, hi-tech industries and uses natural, human and cultural resources already present in a region, but without careful management it can ruin the very resources it is based on.

Switching on and switching off: winners and losers

Developing regions often depend on commodity exports for their income. Bananas are a good example. There are two types:

- 'colonial' bananas — from small farms in former French, British and Dutch colonies in Africa and the Caribbean
- 'dollar' bananas — large plantation-grown bananas from Latin America sold by American TNCs (Dole, Del Monte and Chiquita)

In 1996 Banana Wars began because the EU favoured bananas from the small colonial producers and applied a 20% tariff to banana imports from dollar banana exporters. The UK and France argued that supporting small developing world producers was crucial to the economies of those countries. The USA, dollar banana countries and TNCs that control 70% of the dollar banana trade disagreed. A WTO ruling in April 2008 backed the dollar banana producers. In 2011, the EU agreed to reduce import tariffs on dollar bananas.

This example shows that traditional connections based on links between nations are being replaced by connections created by TNCs. The Banana Wars have both winners and losers:

- Consumers in the EU are likely to get cheaper, better-quality, dollar bananas.
- The Big Three banana TNCs will extend their sales and profits.
- The dollar banana countries will see trade grow.
- The ex-colonial countries and small farmers are likely to see shrinking trade and income.
- For workers on the dollar plantations, jobs will continue, but pressure groups argue that working conditions and wages are worse than in the colonial banana countries.

Knowledge check 65

Outline the likely winners and losers in the Banana Wars.

Switched off: the global periphery

Some peripheral locations have been bypassed by globalisation and remain switched off and seemingly trapped in poverty. Sub-Saharan Africa and parts of south Asia, central South America and the near east all fall into this category. The periphery has not attracted investment for a number of reasons:

- **Political instability:** wars, conflicts, poor governance and corruption deter investors due to high risks.
- **Debt** stifles economic growth and prevents investment in human resources through provision of education and health.
- **Lack of a good infrastructure** — efficient roads, rail, power and water supplies — means production costs are high and distribution is unreliable.
- **Physical constraints**, such as a harsh terrain and unreliable climate, combine with **environmental hazards** such as flooding and soil erosion.
- **Poverty and disease** produce a low-skilled, inefficient workforce that would cost too much to train.

The combination of these factors creates marginalisation. Where investment in the periphery does occur, through tourism and resource exploitation (mining, oil drilling), it is highly localised, creates few jobs and fails to produce connections that might generate broader economic growth.

Roots

Changes in UK population

Over the last decade, net migration (the difference between immigration and emigration) into the UK has been positive. This is unusual, as net emigration was the dominant trend for much of the twentieth century. The population is also ageing. The proportion of people over 60 is expected to rise to 26% by 2020 and 38% by 2050. The twin challenges of managing migration and a greying population are key issues for the UK.

Table 10 The UK's changing population

Statistic	UK figure (2006)	Trends	Explanation
Fertility rate (per woman)	1.84	Declining from 3.5 in 1900 to a low of 1.6 in 2001. Since 2001 there has been a slight upward trend.	Women marrying and having children later; more career focused. Children seen as a cost. Availability of contraception and abortion.
Birth rate (per 1,000)	10.7	Fallen from 28/1,000 in 1900. After the Second World War birth rates rose — the so-called 'baby boomer' generation.	Recent migration has helped the birth rate begin to rise; 20% of all UK births are now to migrants.
Death rate (per 1,000)	10.1	Fallen from 16/1,000 in 1900, now steady. Only the two world wars caused the trend to briefly reverse.	Improved healthcare, diet and education have driven the death rate down.
Infant mortality (per 1,000)	5	Dramatic fall from 140/1,000 in 1900	Improved medical/NHS postnatal care since 1945.
Life expectancy (years)	81 for women 76 for men	Rising from 49 for women and 45 for men in 1900. The difference between men and women has existed for centuries.	Post-war vaccination programmes eradicated diseases such as polio and TB; housing has dramatically improved.
Population age structure	Average age 39	Risen from 34 in 1971. By 2015, the number of people in the UK over 65 will exceed those under 16 for the first time.	Life expectancy improvements have driven up average age.
Family size	2.4	Fallen from around 6.0 in 1900. 29% of UK households are one person, and 36% two person. The most common family unit is a couple with two children, but this is becoming less common.	Fertility decline has contributed to the fall as has the rise in single people and people who never marry. The rise in divorce has created many smaller households.
Ethnicity	8%	Grown from a tiny proportion in 1945. Around 50% are Asian/Asian British, with black/black British another 25%.	Post-war migration is the main driving force. This began in the 1950s and has continued since.
Total population	60.5 million	Total population in 1900 was 38 million. Population continued to grow steadily until 1971 (56 million) but since then growth has been slow.	Lower fertility from the 1960s led to population stagnation in the 1970s. Recently population has begun to grow steadily again, partly as a result of increased immigration.

Researching your 'roots'

Either as a personal or class activity, you should research your own roots, as follows:

- Begin by profiling your own surname using: **www.nationaltrustnames.org.uk/**
- Conduct interviews with family members, focusing on where your family lived in the past and their employment history.
- Consider plotting changing family size over three or four generations.
- Compare your results with those of your peers to identify contrasts between urban and rural roots, and different ethnic or religious roots.

Employment and social status

One of the most significant changes in the UK has been the decline in primary and secondary employment, and the industrial transition. In 1920 there were 1.2 million coal miners in the UK; today there are just a few thousand. Even in the mid-1960s 1.5 million people were employed in the textile industry. Today 76% of the workforce is employed in the service industry, rising from 58% in 1975.

The transition to a service-based economy has been painful. In 1983 there were 3 million people unemployed (12% of the workforce) as the UK deindustrialised. By the twenty-first century the tertiary economy had created many new jobs, including highly paid professional jobs, and quaternary industry had grown. Only 10% of people owned their own home in 1900, compared with 68% today. Rising affluence has led to mass car ownership, foreign holidays and second homes. The sum of all of these changes has been to extend social mobility and increase the proportion of people in the UK who consider themselves 'middle class'.

Knowledge check 67

What is meant by the term deindustrialisation?

UK population geography

Deindustrialisation and the move to the service industries has created considerable regional population change:

- internal migration to the south and east, averaging 30,000 people per year since the 1970s
- rural depopulation, particularly in more remote rural areas, as agriculture has mechanised
- a movement of people out of declining city centres, and into suburbs and the rural–urban fringe
- counter-urbanisation from cities to the accessible countryside, averaging about 90,000 people per year since the 1970s
- international immigration into cities, especially in the south, midlands and northwest

The greying population

The ageing, or greying, population has its roots in the 'baby boom' of 1945–65, when birth rates were higher than today. People born in 1945 reached age 65 in 2010. Those born in 1965 will be 65 in 2030, when 25% of the UK's population will be over 65.

The ageing issue has been made more acute by 'Generation X', born in the 'baby bust' era (1965 onwards) as fertility rates fell dramatically due to female emancipation and the availability of the contraceptive pill. Generation X will be the working-age population in 2010–30 required to support the greying population (Figure 27).

In the future, **dependency ratios** (the ratio of over 65s and under 16s to the working-age population) will rise. In 1971 there were 3.6 working-age people to every pensioner, falling to 3.3 in 2003 and projected to fall to 2.3 by 2051. The ageing population presents us with a range of challenges and opportunities (Table 11).

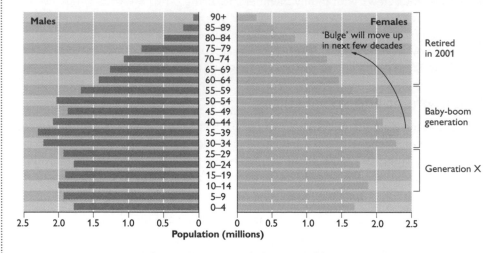

Figure 27 UK population pyramid, 2003

Table 11 The greying population: challenges and opportunities

	Challenges	Opportunities
Economic	Maximising employment (80%+) and tax revenue. The costs of providing long-term care will rise. NHS costs will increase, especially for expensive long-term chronic conditions. General taxation may need to rise. Skilled labour shortages may occur.	Many older people do not want to retire at 65; allowing them to work would increase tax take. A system of personal pensions, for those who can afford it, would relieve pressure on the state system. Older people have valuable experience. The 'grey pound' is a source of economic growth and new market opportunities. Many OAPs have money to spend on age-specific products, from Saga holidays to stairlifts.
Social	Housing will remain occupied, so new housing needed for younger people. The number of single-person retired households will rise. Pressure to raise the state pension retirement age will grow.	An older society may be a more law-abiding one: less need for police and prisons. A desire to 'do good' and 'stay active' may lead to a rise in voluntary and community work. Older grandparents, still healthy and fit at 70 or 80, may take more responsibility for childcare.
Environmental	New sites will need to be found for care homes and retirement homes. Falling population could cause an increase in depopulation and dereliction in some areas.	The 'grey vote' may become a powerful political force, possibly focusing on issues such as equality and environmental sustainability.

The UK may have to consider some radical actions in order to offset the ageing population:
- raising the retirement age
- using the tax system to encourage private pension provision
- using immigration to help raise fertility and avoid population decline
- changing attitudes so that older people are seen as assets rather than costs

Knowledge check 69

What increased costs will governments face in the future as the UK population continues to age?

On the move

In our globalised world, migration has increased and millions of people every year tear up their roots and move, legally or illegally, to new locations. This 'demographic globalisation' is a challenge for both source (origin) and host (destination) countries. In the UK, migration has increased the diversity of the population. If a country receives 500,000 immigrants in a year, but loses 300,000 emigrants, then the net immigration for that year is 200,000. Historically the UK is a country of net emigration (Figure 28).

Figure 28 UK net migration balance since 1901

Examiner tip

Describing the trend of a graph is an important exam skill. The trend in Figure 28 is upwards, but there is also a repeated cycle pattern swinging from net emigration to net immigration. Try to quote precise data from graphs and tables as part of your answer.

In the 1950s and 1960s the UK encouraged immigration from colonies and former colonies such as the West Indies, India, Pakistan and Bangladesh (Commonwealth countries). This met an economic need for low-wage workers and explains the current concentration of these ethnic groups in northern industrial cities (Leeds, Manchester, Bradford) and London. A combination of government policy, lack of immigration barriers, family ties, English speaking and available low-skill jobs maintained the post-colonial flow.

Knowledge check 70

What factors encouraged post-colonial immigration to the UK after the Second World War?

Meeting its international obligations, the UK also accepted refugees from war-torn parts of the world, including:
- 30,000 Asian Ugandans in 1972
- 20,000 Vietnamese boat-people in the 1970s
- 8,000 Bosnian refugees in the early 1990s
- 100,000 Somalis fleeing ongoing conflict since the early 1990s

Some refugees return home, but many cannot. Those who stay may apply for asylum and eventually become citizens.

Immigration has not been evenly spread across the UK. Some regions are much more diverse than others e.g. London is more multi-ethnic than the northeast of England. Ethnic groups have concentrated in particular locations, such as Pakistanis in Yorkshire, Humberside and the West Midlands, and black Africans in London.

Immigration has a role in total population numbers because:
- population numbers rise immediately if immigration exceeds emigration
- migrants tend to be young and have higher fertility, boosting population by higher birth rates (some migrant's cultures value larger families)

In the twenty-first century, immigration has become a political 'hot potato'. There are several reasons for this:
- The number of people seeking asylum in the UK rose sharply from the late 1990s, peaking at 90,000 in 2002. Numbers had fallen by 75% by 2007 because of much tighter government policy.
- Immigration from the eight new eastern European EU member states since 2004 has resulted in the largest mass immigration in the UK's history. The number of eastern European migrants in 2008 is estimated to be around 1 million.
- There are growing concerns about the number of illegal migrants entering the EU (and ultimately the UK) across the 'porous' EU borders along the Mediterranean and in eastern Europe.

New migrant flows have created a sense of 'migration churning', which many people are uncomfortable with. This churning makes planning services such as schools, housing and hospitals challenging. The big question is whether, on balance, migration is a cost or a benefit.

Compulsory case study eastern European migration

Eight eastern European member states acceded to the EU on 1 January 2004: the Czech Republic, Estonia, Hungary, Latvia, Lithuania, Poland, Slovenia and Slovakia (the A8 or Accession Eight countries). The EU has an 'open border' policy for both trade and people. The 1995 Schengen Agreement removed border controls between many European countries. Migration to the UK from A8 countries has been on a much bigger scale than that to all other EU countries. In 2003 government estimates of the number of migrants who might move to the UK in 2004–05 were put at 20,000–40,000. In fact, over 200,000 immigrants arrived. This is explained by a number of **pull factors**:
- In 2004, A8 countries were not given the right of free movement except by the UK, Ireland and Sweden (the UK is not part of the Schengen Agreement and retains border controls), which narrowed the choices for potential migrants.
- The developed economy of the UK is a magnet for migrants seeking work and opportunities.
- The UK has a reputation for tolerance towards migrants.
- A growing, low-unemployment economy was in need of low-wage labour.
- The intervening obstacles for new migrants were relatively minor because:
- a Worker Registration Scheme allowed migrants right of entry and the right to work
- cheap bus and air travel reduced costs

Knowledge check 71

Explain two ways in which immigration causes total population numbers in a country to rise.

Examiner tip

Be clear about what is meant by 'source country' and 'host country' as these terms can get confused in the exam. Migrants originate from a source country (their home country or origin) and move to a host country (their destination).

Knowledge check 72

Name the A8 countries and state the date when the group of countries joined the EU.

- existing established immigrant groups, such as the Poles, provided community and cultural support for newcomers
- immigrant-friendly services, such as 'Polish' UK bank accounts, were quickly set up, allowing remittances to be sent home

The number of migrants grew quickly from almost none in January 2004 to 665,000 resident migrants by late 2007. Most of the migrants work in low-paid temporary jobs in agriculture or services such as catering. They are spread across the UK, including rural areas such as the Scottish Highlands and Herefordshire. Table 12 outlines the key costs and benefits of A8 migration.

Knowledge check 73

State the economic costs and benefits to A8 source countries of emigration to the UK.

Table 12 Costs and benefits to the UK and A8 countries of post-accession migration

	Source countries (A8)	Host country (UK)
Costs	Loss of workforce: 7% in Poland and 10% in Lithuania. Brain drain and skills shortages: skilled migrants such as plumbers and doctors have left Poland. Ageing and reduced fertility at home; Poland's population could decline from 38 to 33 million by 2050. Most migrants are young males creating an imbalance at home. Some workers exploited by gang-masters and are paid less than minimum wage.	Social/cultural tensions, especially in rural areas with no history of immigration. A rise in low-level 'cultural unfamiliarity' crime, e.g. road traffic offences. UK population growth to 65 million by 2016; 20% of babies are born to recent migrants (8% of the general population). Downward pressure on wages as A8 migrants 'undercut' low-wage Britons. Pressure on space and housing causing locally rising housing costs. Pressure on local schools, the NHS and council services.
Benefits	Remittances: workers send 25% of earnings home. For Poland this meant €6.4 billion in 2006. Higher wages, A8 migrants average £6 per hour, higher than at home. Skills can be taken back, e.g. English and IT skills. Working A8 migrants abroad do not need benefit payments.	Skills gaps filled in industries, e.g. fish processing in the Highlands and farming in the Fens. Economic turnaround in areas such as South Lincolnshire where 10% of the population are A8 migrants, meaning more local spending. Business opportunities for banks and supermarkets providing Polish food and services. Counteracts ageing if migrants stay and have children.

Approximately 1 million A8 migrants have come to the UK since 2004, but around 350,000 have returned home, illustrating that much of this migration is temporary economic migration.

The future

Eastern European migration to the UK has slowed since 2008 because:

- A8 economies like Poland were less badly hit by the global recession than the UK. In the UK unemployment increased in 2008–2010 so there were fewer job opportunities for migrants.
- The falling value of the UK pound has reduced the value of remittances sent home by A8 workers.
- In 2011, all EU countries must be fully open to A8 migration and many migrants will 'divert' to other countries.

The A2 countries, Romania and Bulgaria, which joined the EU in 2007, had restrictions placed on free migration. These restrictions will be gradually lifted up to 2013.

Examiner tip

You will need to learn the compulsory migration case studies in detail and be prepared to answer shorter Section A and longer Section B questions that refer to them directly.

Compulsory case study **Mediterranean flows**

UK nationals increasingly emigrate south, towards sunshine and warmer temperatures. Around 750,000 Britons live permanently in Spain, and 200,000 in France. Total emigration reached 400,000 in 2008; approximately half of these emigrants were British (the rest were returning immigrants). These 'sunseekers' are very different from the A8 migrants. Most of them (60%) are economically inactive. Many are retired and some own businesses such as hotels.

Migrants to the Mediterranean are pulled there by a range of factors:
- Housing, and the cost of living, is generally cheaper than in the UK.
- As a growing tourist destination, the Med has many new business opportunities.
- Better weather is guaranteed — hot summers and mild winters.
- Lifestyle may be more relaxed, with a 'holiday atmosphere'.

There are also push factors. Some are simply the reverse of the pull factors above (bad weather!) but many migrants cite factors in the UK such as:
- a perception of rising crime, declining respect and a sense that the UK has 'gone to the dogs'
- high taxation and spiralling house prices
- lack of space, congestion and low environmental quality

Retirement migrants may be able to sell a valuable house in the UK, buy a much cheaper one in Spain and have a substantial sum left over to add to their state/ personal pension.

Other developments have made the process of emigration easier since the 1980s:
- Spain joined the EU in 1986.
- The Schengen Agreement and the 1992 Maastricht Treaty (which created the Eurozone) have made moving to another EU country much easier.
- Low-cost airlines provide easy and cheap access.
- Companies employing English-speaking lawyers and estate agents can sort out legal issues and paperwork.
- The internet allows easy communication with family members, friends and business interests back in the UK.

A criticism of 'sunseekers' is that they are not integrated with local communities and culture. This has intensified over time as English bars, doctors, schools, sports clubs etc. have set up, so there is even less need for emigrants to integrate with their Spanish or French hosts. There are extreme cases, such as San Fulgencio on the Costa Blanca: of its 10,000 population, 8,000 are foreign (half of these are British). The area was in the news in 2007 as British and German expats set up their own political party and stood in local elections, winning 21% of the vote and three local council seats.

This type of migration has slowed since 2008. Sunseekers, especially pensioners, are sensitive to the value of the UK pound. As this has fallen, so have their incomes. Some have returned to the UK. Others feel trapped abroad as the value of their homes has fallen, making it hard to sell up and return to the UK. Table 13 outlines the key costs and benefits of retirement migration to the Mediterranean.

Knowledge check 74

State the factors that have made it easier for UK emigrants to move to Spain and other Mediterranean locations.

Examiner tip

It is important to discuss the costs and benefits of migration in a factual, balanced way in the exam. Make sure your answers are based on sound geography and not tabloid newspaper headlines.

Table 13 Costs and benefits of retirement migration to the Mediterranean

	Source country (UK)	**Host country (e.g. Spain)**
Costs	**Social** Family breakup, as grandparents move away. Loss of potential childcare. **Economic** Loss of a highly experienced workforce, especially if they retire early. The 'grey pound' is spent overseas.	**Social** Emigrant 'ghettos' are created, with little social and cultural integration. Resentment may grow as immigrants seek to enter local politics. House prices exceed the buying power of local people. **Economic** Some benefits and healthcare costs are borne by the host country. **Environmental** Large-scale villa development has ruined much of the coastal landscape and degraded biodiversity. Water supply systems are strained in semi-arid areas. Localised pollution has risen, and flood risk rises as urban development occurs.
Benefits	**Social** In part, emigration balances increased immigration, reducing net migration rates. **Economic** Fewer older people to take care of; some health and care problems are effectively exported. **Environmental** Relieves pressure for new homes, and therefore to build on greenfield sites.	**Economic** Increased spending in the local economy. Some retirees are highly affluent. Job creation in construction, retail, legal and health services. Areas which were largely unproductive scrubland become valuable land to build on.

Managing migration

Managing migration is a key issue for governments. Globalisation has pushed it to the top of most political agendas in the developed world. Migration policy is a tricky balancing act, as Table 14 shows.

Table 14 Managing migration: policies

Policy	Advantages	Disadvantages
Border controls	Physical borders, policing, passports and visas allow governments to count people in and out and so keep track of net migration.	These systems are extremely costly (the UK plans to introduce 'electronic' border control in 2010, at a cost of £2 billion). Tight regulations may put off some visitors and migrants who are needed to fill skills gaps.
Work permits	Work permit schemes allow temporary workers to be controlled and matched to skills shortages.	Work permits can be abused, with some not leaving when their 'time is up' and becoming illegal migrants.
Refugees and asylum seekers	Prestige is gained by accepting vulnerable groups and respecting basic human rights.	The public may perceive refugees and asylum seekers as a cost with few benefits attached. The asylum system is costly to administer.
Integration	Citizenship tests such as those in the UK and USA might help integration by expecting a basic understanding of language and cultural norms.	Critics argue that passing a test does not prevent social tensions and racism.

Examiner tip

Migration is a dynamic topic that does need updating year on year as government policy changes, flows of migrants shift and net migration rises and falls. Make sure you keep up to date.

Knowledge check 75

What attracted many retired people from the UK to move to Spain?

In the UK, migration policy has tightened recently as public concerns about mass migration have risen. UK government policy has moved towards an Australian-style points-based immigration system. The system needs to balance the economic need for skilled workers and younger people to counteract the greying population, while minimising the social and environmental costs of immigration.

World cities

A total of 3.3 billion people live in urban areas. In 2007–08 the number of urban dwellers exceeded the rural population for the first time.

- In 1900, 10% of people lived in cities; by 2030 this figure will rise to 60% of the global population and by 2050 it will be 75%.
- Half of the urban population today is under 25.
- Asia's urban population is set to rise from 1.4 billion to 2.6 billion by 2030.
- Africa's urban population will rise from 300 million to 750 million, and Latin America's from 400 million to 600 million.
- Some cities are considerably larger than major countries: Greater Tokyo (35 million) is larger in terms of population than Canada (33 million).

> **Knowledge check 76**
>
> Why was 2007–08 a very significant year for global urbanisation trends?

The definition problem

Deciding what constitutes a 'city' is far from straightforward. Even for a small provincial city such as Leicester there are problems. In 2006, 290,000 people lived within the city's administrative boundary, but some suburbs are in other administrative areas, so in fact the urban area had a total population of 440,000. One of the problems with population figures is that most of them are estimates, as censuses are carried out only every 10 years, and even they 'miss' some people (illegal workers, temporary migrants).

Urban agglomerations and megacities

> **Examiner tip**
>
> Megacities are classified by size (population over 10 million) whereas world city status is based on economic and political factors. Some cities, such as New York, are megacities as well as world cities.

It may be useful to consider urban areas and wider metropolitan areas, rather than cities. **Urban agglomerations** — similar to **conurbations** — exist when urban areas merge with each other. Urban areas with a population in excess of 10 million are defined as **megacities**. There were 20 megacities in the world in 2007, 15 of which were in the developing world. Many are expected to grow to reach 20 million people by 2015. New cities will also acquire megacity status. By 2020, or very soon after, there could be close to 30 megacities around the world.

There are 200+ cities with populations of over 1 million — around 70 of these are in China. These 'million cities' include many cities in the developed world that are stagnating or declining. A few lucky cities have attained 'world city' status. These cities have significant global economic and political power:

- **London** — global financial centre, world's busiest international airports, 23 Global 500 companies
- **New York** — global financial centre, 24 Global 500 companies, home of UN
- **Tokyo** — global financial centre, 52 Global 500 companies

Increasingly, megacities are transforming themselves into new urban forms. As cities grow, they reach a size limit (approximately 15 million people) beyond which the city cannot grow with one central business district (CBD). As this limit is reached the city becomes so congested that people begin to move outwards. Cities become more linear and a string of urban centres grows into a **megalopolis** or '**urban archipelago**'.

It is important to note that the rate of urbanisation is highly variable across the planet, with some cities growing at breakneck speed and others slowing to a crawl. There is a contrast in the nature of urban processes in large cities, depending on their location in the world and stage of economic development (Table 15).

Knowledge check 77

What is the difference between a world city and a megacity?

Table 15 World urbanisation types

	Urbanisation stage	Growth	Economy	Planning and cycle of urbanisation	Examples
Developing world	Immature	Very rapid: 3%+ per year. Largely migration growth.	Informal economy = 60%. Small-scale manufacturing, street trading, urban farming.	**Urbanisation** Little planning: uncontrolled sprawl. Squatter settlements dominate. Basic needs barely met. 60%+ live in slums. Environmental problems.	Kabul Lagos Kinshasa
	Consolidating	Rapid 2–3% per year. Balance of migration and internal growth.	Manufacturing important, some service industries. Informal economy around 50%.	**Urbanisation and suburbanisation** Attempts at planning, focused on waste, congestion and water supply. Upgrading of slums and some social housing. Most basic needs met.	Cairo Mumbai Jakarta Chongqing
	Maturing	Slow: under 2% per year. Largely internal growth.	Service industry dominates, with some manufacturing. Informal economy under 40%.	**Suburbanisation** Effective attempts at housing, transport and land-use planning. Environmental problems being tackled. Quality of life satisfactory for many. Gated communities in suburbs.	Mexico City São Paulo Beijing
Developed world	Established	Very slow, under 1% per year. Some are stable.	Dominated by professional, services and retail. Formal economy.	**Counter-urbanisation and reurbanisation** Large-scale suburbanisation, with counter-urbanisation. Since 1980, most have regenerated inner-city and former industrial areas. Quality of life is high for most, and environmental quality is good.	London San Francisco Paris Birmingham

Urban growth

The world's most rapidly urbanising cities have high population growth rates. Much of this is fuelled by rural–urban migration. This is most acute where rural poverty is highest. Food shortages, lack of land due to subdivision, fuel shortages, soil erosion, war and conflict, natural hazards — all push people out of rural areas. Perceived opportunities — jobs, better housing, healthcare, a chance of an education — are the pull factors.

Rural–urban migration has a double effect. Most migrants are young and fertile. This creates high natural increase, further fuelling population growth. In some developing cities, natural increase and rural–urban migration are contributing to staggering rates of urban growth. By 2015, every hour 42 people will be added to the population of Mumbai, 50 to Dhaka and 58 to Lagos.

Knowledge check 78

What are the two processes that lead to urbanisation and the growth of urban areas?

In the developed world, trends are widely different:
- Greying populations have reduced natural increase, or even reversed it.

- Urban growth in cities like London is largely due to international immigration.
- Despite regeneration, suburbanisation and counter-urbanisation continue to draw people out of cities.

Because cities are at different stages of the cycle of urbanisation, life can vary considerably from one city to the next.

Rapid urban growth in the developing world has led to a slum crisis. Lacking jobs and forced to work in the informal economy, people build their own settlements. Most lack tenure (a requirement for improvements), and sanitation and clean water are patchy or absent. In Mumbai, for example, 60% of the population live in slums on 6% of the city's land. With a growth rate of 2.2% per year, the slum population will rise from 1 billion in 2005 to 1.4 billion by 2020.

Newcomers to developing world megacities inhabit 'old' slums such as Dharavi in Mumbai. These are tenement blocks and shanty towns with appalling overcrowding and sanitation. They have the advantage of being close to work in the city. Squatter settlements spring up on the 'septic fringe' through organised landgrabs. Slums are located where no one else wants to be — on dangerously steep slopes, next to polluted rivers, on marshland or downwind of polluting industry.

In both the developed and developing world the disadvantages of the inner city have created 'movers' who increasingly suburbanise to the urban edge. New suburbs are often 'gated' (i.e. walled off with security gates). This trend increases the segregation between rich and poor.

Managing cities

Megacities face a number of challenges depending on their stage in the cycle of urbanisation. In the developing world, problems usually result from a lack of resources; in the developed world, the problem is often overuse of resources and large urban ecological footprints. Both situations are unsustainable. At both ends of the development spectrum, a lack of management of urban waste and systems leads to environmental problems.

Environmental issues

In developed cities, ways need to be found to reduce inputs, for example through:
- water metering and pipe mending
- reduction in the use of packaging
- more public transport
- more efficient vehicles
- recycling building materials
- localised food distribution and improved storage

Outputs too must be reduced, by:
- using less polluting vehicles
- finding alternative energy sources
- recycling
- water reuse
- more carbon sequestration, e.g. urban gardens, farms and forests

Such methods must be implemented if ecofootprints are to decrease and urban quality of life improve.

Ecocities

China plans to build a series of ecocities to showcase urban sustainability. The first of these, Dongtan, is due to be constructed close to Beijing and could accommodate up to 500,000 people.

- It will be a largely car-free and carbon neutral.
- Energy will be provided from renewable sources — wind farms and solar panels.
- A waste treatment plant will convert sewage and compost into biogas that will be used for cooking, heating and power generation.
- Most people will live in apartment blocks six to eight storeys high, designed with natural ventilation to minimise the need for air conditioning.
- There will be two water-supply systems: one with drinking water and another providing 'grey water' for toilets and garden irrigation.
- Traffic lights will automatically switch to give priority to hydrogen-fuelled buses.

Construction of Dongtan has been delayed. This was partly due to the global recession in 2008–2010. Building new ecocities or 'greening' existing cities is expensive and needs a lot of political will to push through.

Knowledge check 80

Outline the main aims of an ecocity or ecotown.

Improving slums

However, building from scratch in this way is not a viable solution for the millions of slum dwellers in rapidly expanding megacities. Help often comes from NGOs working in slum districts, with funding from local authorities, NGOs themselves and organisations such as the World Bank. Often, improving slums focuses on:

- providing residents with tenure (the right of ownership) as a first stage in giving them confidence to make improvements
- providing basic services, such as clinics and schools, often by city authorities
- NGOs providing improved water supply such as tubewells and standpipes
- residents themselves improving sewers, roads and houses through 'aided self-help' schemes, which provide building materials and basic training

There is a place for large-scale slum clearance and major sewerage and water engineering projects, but in many megacities the scale and continual growth of the slums defy this approach.

Examiner tip

Use the term 'slum' to describe low-quality housing in developing world cities, rather than shanty towns or squatter settlements. The United Nations uses 'slum' to include any low-quality, poorly serviced homes, not just those built by the residents themselves.

Global challenges for the future

The globalised world is increasingly:
- urban
- on the move
- connected

Knowledge check 81

Why is providing slum dwellers in developing world cities with tenure important if slums are to be improved?

This is not the case for everyone, or even for the majority. Globalisation has created more choice but also more homogeneity — a Starbucks and a McDonald's in the centre of every city. Demographic globalisation has created mixing and diversity, but with it come the social and cultural tensions associated with rising migration. We are all global consumers but we live in a world fearing global climate change and worrying about resource shortages. Globalisation has a price. How can it be managed to reduce the costs?

A two-speed world?

Despite the gains from globalisation in countries such as China and India, the world continues to be widely unequal. In 1960 the ratio of the income of the poorest 20% of the world population to the richest 20% was 1:30. By the year 2000 this had risen to 1:70. In many parts of the developing world, the poor have been getting wealthier, but developed countries have been getting wealthier faster, thereby widening the gap. In many African countries, income levels have actually fallen since 1980.

Dramatic rises in consumption of resources are explained by global population growth (from 3.7 billion in 1970 to 6.4 billion in 2005) and higher consumption per person. Chinese consumption has risen dramatically and by 2031 may reach levels similar to those of the USA in 2005:

- 1,350 million tonnes of grain per year (66% of world production in 2005)
- 180 million tonnes of meat per year (80% of world production in 2005)
- 2.8 billion tonnes of coal per year (global production 2.6 in 2005)
- 1.1 billion cars (800 million globally in 2005)

Goods that were once sourced locally, or were unavailable, are now imported by developed countries from overseas. This issue of 'food miles' — the distance that produce travels from farm to consumer — is important. The longer the distance, the bigger the need for refrigeration and for extra packaging, and the higher the 'carbon footprint' of that particular food. Some examples in the UK are:

- grapes from Chile — a distance of over 12,000 km, heavily packaged to avoid damage
- strawberries from Spain — a distance of approximately 1,600 km
- prawns from Indonesia — a distance of almost 12,000 km
- sending frozen Scottish-caught prawns to China for processing into scampi, then shipping these back to the UK for consumption
- sending coffee produced in Africa to India for roasting, then to the UK for sale

Exploitation or development?

The 'winners' from globalisation might be expected to be those that have gained jobs as TNCs have shifted production to low-cost locations — but is this true?

A typical factory worker in China's Pearl River Delta has the following working conditions:

- Pay is $50–150 per month.
- Unions are banned.
- Overtime, a requirement to keep the job, is around 30 hours per week.
- Many migrant workers live in factory-owned dormitories, 16 people per room.
- Labour and health and safety laws are rarely enforced.

The alternative work is farming, which in China nets an annual income of $300–400 compared with $1,000+ for a factory worker. The majority of factory workers are rural–urban migrants who consider their lives improved compared with those of their parents. There are moves to increase worker rights and ban child labour in India and China. For governments there is a difficult balancing act as attempts to improve workers' conditions could simply mean TNCs shifting again, to the next low-cost location.

Knowledge check 82

What is the main concern about 'food miles'?

Examiner tip

Many factory workers in Asia have poor working conditions. This raises ethical issues when we buy the products they make. Be careful not to imply that these factory jobs are equivalent to slave, bonded or child labour — all of which are illegal.

Knowledge check 83

Working conditions in the factories of China's Pearl River Delta area are hard, so why have so many migrants moved there?

Green globalisation?

If a globalised world is to achieve sustainability, action is needed at a range of scales. There is a hierarchy of possible actions:

- individual actions — changing behaviour
- local actions — usually by local councils
- national — government policies
- global — agreements and targets

Concern over rising carbon emissions has moved up the political agenda and is now on the radar of many consumers. Those seeking to reduce their carbon footprint have a range of options (Figure 29).

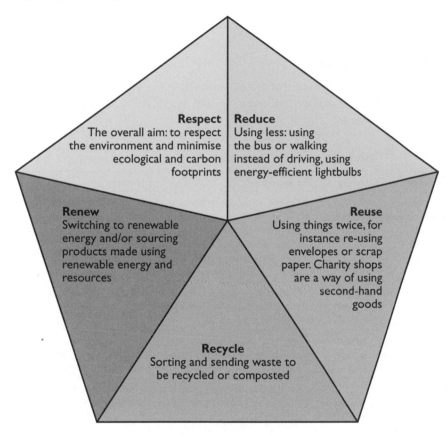

Respect
The overall aim: to respect the environment and minimise ecological and carbon footprints

Reduce
Using less: using the bus or walking instead of driving, using energy-efficient lightbulbs

Renew
Switching to renewable energy and/or sourcing products made using renewable energy and resources

Reuse
Using things twice, for instance re-using envelopes or scrap paper. Charity shops are a way of using second-hand goods

Recycle
Sorting and sending waste to be recycled or composted

Figure 29 The 5 Rs

Individuals can also 'offset' their emissions by paying for new tree planting to sequester their pollution. Many airlines offer this service on their websites — guilt-free global air travel — but is it?

Governments can change consumer behaviour. Car tax (Vehicle Excise Duty, VED) is one of the UK government's ways of applying a 'green tax'. It aims to use the tax system to change our car-buying behaviour by linking VED to carbon dioxide emissions per kilometre. Introduced in 1998, variable VED does seem to have affected the types of car we buy.

Knowledge check 84

If you wanted to buy a car with low VED (annual car tax), what type of car would you buy?

Many businesses, even TNCs, have jumped on the 'green' bandwagon as public concern has grown. In 2002 BP changed its company logo from a shield to a flower, and the company has diversified into solar power and retailing. Many people are cynical about the motives of business for 'going green' but there are reasons to suggest companies such as BP could be serious:

- Money can be saved by reducing energy use (and therefore pollution).
- Company assets are vulnerable to sea-level rise and flooding linked to climate change.
- Oil and gas will run out; renewable energy provides an alternative business model.
- Consumers are increasingly 'switched off' by companies with a poor environmental record.

If companies can see long-term benefits to shareholders of 'going green', they are likely to move in this direction.

Ethical globalisation?

Ethical concerns over the nature of globalisation have led some consumers to change their behaviour in the hope of reducing the negative social consequences of consumption (Table 16).

Examiner tip

Think critically about the costs and benefits of ethical schemes such as fair trade. These often work well and help provide developing world workers with their basic needs. However, fair trade farmers gain only 10–15% more income than the alternative, which provides only limited capacity to improve lives.

Knowledge check 85

How does fair trade help farmers in the developing world?

Table 16 Taking action to create a more equitable world

Strategy and benefit	Example	Issues
Fair trade Developing world farmers get a higher price for their products.	Coffee, chocolate, bananas and some clothing. In 2010, the total value of UK fair trade sales was £1.1 billion and it is growing at 40% per year.	Hard to monitor the growing number of schemes. The fair trade price premium is small.
Ethically sourced goods Guaranteed free of child or 'sweatshop' labour.	Many shops and supermarkets have ethical clothing ranges.	Currently only a small % of all goods sold.
Buying locally Reduces food miles and supports local producers.	Seasonal 'veggie boxes' are delivered by farms to many households in the UK.	More expensive than using supermarkets; reduced choice.
Boycotts Consumers put pressure on businesses with poor ethical or environmental records.	Nestle has been the subject of boycotts over its formula baby milk.	May have little impact unless it becomes very widespread.
Organic food Reduces the carbon footprint of food and pollution from farm chemicals.	Many supermarkets now sell organic milk, meat, cheese, fruit, vegetables and cereals.	Often more expensive than non-organic food, with less choice.
FSC certified Ensures that wood products have been produced from sustainable forests.	Hardwood garden furniture often has an FSC mark.	Difficult to monitor especially from distant sources in the developing world.

Going global

Summary

- Globalisation is a process which leads to the growing interdependence of countries worldwide.
- The drivers of globalisation include communication and transport technology, the development of TNCs and the removal of trade barriers.
- There are growing global networks of trade, communications, transport and forums such as the UN for political negotiation, all leading to shrinkage of space and time.
- Globalisation is fundamentally an economic process, but includes cultural, political, demographic and environmental dimensions.
- There are both positive and negative outcomes of globalisation. The global shift of industry and services has created increased consumer choice, with lower priced goods, and the gains in jobs and opportunities have led to poverty reduction in developing nations, but have also led to sweatshops and job losses in developed nations.
- The world is composed of a number of economic groupings based on level of development and wealth (LDCs, LEDCs, RICs, NICs, MEDCs). Groups of nations form trade blocs such as the EU or ASEAN. There are also political groupings such as OPEC.
- Globalisation leads to winners in the core areas such as global hubs and also losers in the switched off peripheral world — creating a two-speed world.
- Flows of economic migrants are attracted by economic opportunities (east Europeans to the UK) and opportunities for lifestyle improvement (retirement migration to the Mediterranean).
- Migration has positives and negatives for both source and host countries, and the management of migration flows and the resultant multi-racial, multicultural society is a complex task for governments.
- A study of roots in the UK shows the pattern of flows over time from the post colonial era to the present day and shows how society has changed.
- As economic development takes place, internal migration has led to rapid urbanisation, concentrated in developing nations.
- The number of world cities and megacities has increased especially in the developing world. Megacities can be classified according to their stage in the cycle of urbanisation. Their existence provides challenges to develop more sustainable cities.
- The increasingly urbanised, connected, mobile world poses challenges for the future. There is a need for a globalised world to become a more equitable world and achieve greater sustainability in terms of resource usage.

Questions & Answers

In this section of the guide, there are seven questions — four Section A questions and three Section B questions.

The Section A questions are on global hazard trends, global challenges for the future, coping with climate change and on the move. Each question is accompanied by answers at different grades and comments on how marks were awarded. There is also some additional exam technique guidance on how to improve your performance in the exam.

The Section B questions are on the impacts of global warming, globalisation and global hazard patterns. Each is accompanied by examiner's comments that identify the good and less good aspects of each answer. Sample A-grade to D-grade answers are also included so you can compare them.

Assessment

In the examination Section A tests the *breadth* of your knowledge and understanding across the whole unit. It is marked out of 65 marks and you have to complete six short structured questions in under 1 hour. A typical question includes, in increasing order of difficulty:

- definition of a term or multiple choice (1 mark)
- description of a resource, e.g. trends, patterns (2 marks)
- development of a resource, by suggesting reasons (3–4 marks)
- explanation question which is not linked to a resource (4 marks)

Questions in Section A are marked in points.

Section B tests the *depth* of your knowledge and understanding. You have to choose one question from the four set, worth 25 marks, and you have around 30 minutes to answer it. Each question consists of two parts:

- data stimulus (10 marks)
- extended writing, using your compulsory case studies (15 marks)

Section B questions are marked using levels, and the quality of written communication is assessed.

World at risk		Section A			Going global
Q1	Q2	Q3	Q4	Q5	Q6
11 marks	10 marks	12 marks	11 marks	11 marks	10 marks

Section A

Increasing difficulty

Typical question
- Definition of a term (1 mark)
- Multiple-choice question (1 mark)
- Simple analysis of a resource (data-response) (2 marks)
- Development of a question on the resource (3 marks)
- Explanation question. Not linked to resource (4 marks)

All are marked in **points**.

Section B

Q1 Hazards (25 marks)

Q2 Global warming (25 marks)

Q3 Globalisation (25 marks)

Q4 Cities and migration (25 marks)

Section B

Each question consists of two parts

(a) Data stimulus, clear structure provided (10 marks)
(b) Extended writing, including using your compulsory case studies (15 marks)

All are marked in **levels**. Quality of written communication is assessed.

Edexcel AS Geography

Section A
Question 1

Global hazard trends

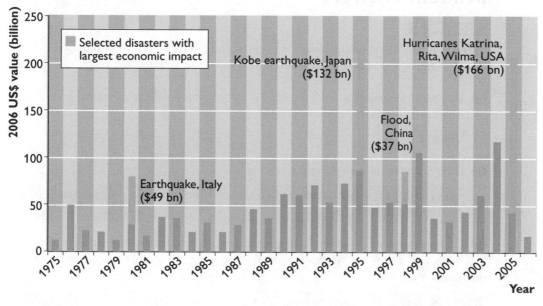

Figure 1 Annual reported economic damages from natural disasters, 1975–2006

(a) (i)   Define the term 'natural disaster'.   (2 marks)

ⓔ To gain both marks, you need to explain both 'natural' and 'disaster'

(ii) Which is the correct statement concerning trends in natural disasters?
  • **A There have been increasing numbers of tectonic disasters in recent years.**
  • **B The number of deaths from disasters has decreased dramatically in recent years.**
  • **C The relative amounts of damage are greater in more developed countries than in developing countries.**
  • **D The number of people affected by natural disasters has increased tremendously in recent years.**   (1 mark)

(iii) Describe the trends shown in Figure 1.   (2 marks)

ⓔ You need two different trends for 2 marks

(b) Give two examples of economic damages.   (2 marks)

ⓔ Each example is worth 1 mark.

(c) Explain, using an example, why certain natural disasters prove to be extremely costly in terms of economic damages. (4 marks)

ⓔ Here you should use detail from a high-magnitude disaster in an MEDC or NIC and focus on a range of economic costs.

Total: 11 marks

Student answers

Student A

(a) (i) A hazardous natural event, e.g. earthquake, **a** that leads to significant damage to people and property. **b**

ⓔ **2/2 marks awarded. a** This is the 'natural' part of the definition. **b** 'damage to people and property' covers the disaster element on the question.

(ii) D **a**

ⓔ **1/1 mark awarded. a** This is the correct answer.

(iii) The trend is generally upward **a** although there are anomalies such as 2006 at about $20 billion. **b**

ⓔ **2/2 marks awarded. a** Identifies the overall trend. **b** Spots an anomaly and identifies it specifically, with use of data. The student could also have stated that there is much variation from year to year/fluctuations.

(b) • Losses to people's property, e.g. homes and businesses.
　　• Crops become inundated and lost during a flood or tsunami. **a**

ⓔ **2/2 marks awarded. a** Both answers are correct. In Section A, using bullets or numbered points is acceptable, so long as they are not too brief, i.e. one or two words only.

(c) Hurricane Katrina caused $80 billion **a** in economic losses because it struck a major city, New Orleans, with many wealthy people in a small, low-lying area. **b** This meant damage to 1000s of homes and businesses. **c** Most were insured, so insurance companies had huge losses. **d** The US government spent billions on aid and the clean up as levées that had been damaged by the storm surge had to be rapidly repaired. **e**

ⓔ **4/4 marks awarded. a** Good example with data on damages, **b** supported by specific points about losses in New Orleans, **c** property losses and **d** explanations of insurance loss. **e** The last point about aid and rebuilding is good, so the answer has the range needed for full marks.

ⓔ **11/11 marks in total.** Grade A awarded.

Student B

(a) (i) This is something that causes major damage to people and their property. **a**

ⓔ I/2 marks awarded. a This is not a full answer as the 'natural hazard' part of 'natural disaster' is not addressed. It is only referred to as 'something'.

(ii) D **a**

ⓔ I/I mark awarded. a This is the correct answer.

(iii) Economic impacts are rising from 1975 to 2005, generally. **a**

ⓔ I/2 marks awarded. a The overall trend is identified, but another point is needed (fluctuations, an anomaly) for full marks.

(b) Examples could be the damage to peoples homes **a** and damage to residential areas. **b**

ⓔ I/2 marks awarded. a The examples given here are basically the same, i.e. homes and **b** residential areas.

(c) Earthquakes cause massive natural disasters and huge economic losses because of severe ground-shaking **a** which flattens buildings and destroys transport networks. **b** All of these are economic.

ⓔ 2/4 marks awarded. a This generalised answer identifies a reason for the losses and **b** states what some losses are, but it lacks use of examples and the idea that some disasters are worse that others due to location or nature of the population.

ⓔ 6/II marks in total. Grade C awarded.

ⓔ Make sure you:
- Pay attention to the mark weightings.
- Always try to use examples to support your statements.
- Always use data from the Figure if you are asked to refer to one.
- Always use correct geographical terminology in your answer.

Question 2

Global challenges for the future

Study Figure 2, which shows ethical behaviour in the UK in 1999 and 2007, and answer the following questions.

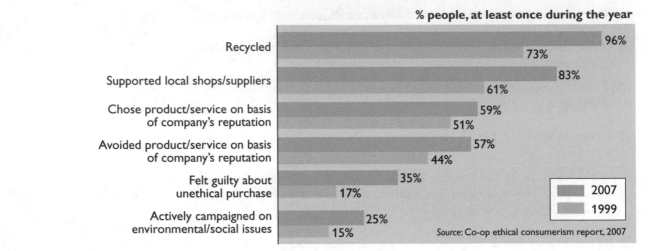

Figure 2 Ethical behaviour in the UK, 1999 and 2007

(a) Which behaviour involves trying to reduce the environmental impact of household waste? **(1 mark)**

(b) Identify two changes in behaviour between 1999 and 2007. **(2 marks)**

(c) Suggest why 35% of consumers in 2007 'felt guilty about an unethical purchase'. **(3 marks)**

ⓔ Although this question does not ask for examples, as a general rule you will gain credit when you include them.

(d) Explain how fair trade might benefit producers in the developing world. **(4 marks)**

ⓔ Focus on fair trade, and make sure your explanations relate to developing world producers (not the consumers of fair trade products).

Total: 10 marks

Student answers

Student A

(a) Recycling a

ⓔ **1/1 mark awarded. a** Correct answer.

(b) Every activity in Figure 2 goes up 1999–2007 **a** but 'felt guilty about ethical purchase' doubles from 17% to 35%. **b**

ⓔ **2/2 marks awarded. a** The overall pattern is stated, **b** plus a specific, quantified change.

(c) Consumers may feel guilty about buying cheap clothes from Primark **a** and other stores as workers' wages are low. Some workers in the developing world only get paid a few dollars per day. **b** People are aware of fair trade more due to advertising. **c**

ⓔ **3/3 marks awarded. a** There is good use of an example, **b** plus an extended point about wages. **c** The last point also scores 1 mark.

(d) Fair trade farmers receive a guaranteed price **a** for their goods, e.g. tea and cocoa, which means they have more money than non-fair trade. **b** They could afford to send a child to school, **c** or invest in better farming methods to increase production and increase income even more. **d**

ⓔ **4/4 marks awarded. a** This answer has a good sequence of explanations beginning with the basic benefit of fair trade, **b c d** followed by three further, linked benefits.

ⓔ **10/10 marks in total.** Grade A awarded.

Student B

(a) Recycled **a**

ⓔ **1/1 mark awarded. a** Correct answer.

(b) People are more likely to recycle than they were in 1999 by 23%. **a**

ⓔ **1/2 marks awarded. a** Although data are used, only one change is identified.

(c) Might have been made by workers who are exploited. **a** Could have caused pollution e.g. water pollution during making it. **b** Food miles. **c**

ⓔ **2/3 marks awarded. a b** The first two points are OK, and each scores a mark but **c** the third point has no hint of a reason for 'guilt' or concern. Make sure your points do not become too brief.

(d) It means farmers get more money than if they did not do fair trade. **a** They could use this extra money for development. **b**

ⓔ **2/4 marks awarded. a** This answer only has two basic points about money and **b** development. The development point gains only 1 mark — it would have been better to state 'education' or 'community projects'. The answer does not pay attention to the mark allocation.

ⓔ **6/10 marks in total.** Grade C awarded.

ⓔ Some Section A questions ask you to include examples whereas others do not. As a general rule, if you include examples you will gain marks. Questions worth 4–5 marks usually use the command word 'explain'. You will need to give a range of reasons/explanations. In the question above on fair trade, most students could state that fair trade increases the incomes of producers. In the exam, when you have written this, ask yourself 'so what?'. In this case the answer to the 'so what?' could be that the extra income allows the producers to buy extra food, so they can work harder because they are healthier. The 'so what?' forces you to think of further reasons and extend your answer.

Question 3

ⓔ In Section A, questions usually end with a 5-mark question which uses the command words 'explain' or 'suggest reasons'. There could be six of these questions in Section A, totalling 30 marks. They are very important in terms of the overall grade you will achieve. The questions are not linked to a Figure, but they are not 'essay' questions. They are point marked. You need to make sure you write enough to achieve 5 marks but not so much that you end up writing an essay and wasting time.

Coping with climate change

Using examples, explain the difference between adaptation and mitigation strategies to deal with global warming.
(5 marks)

ⓔ Note that examples need to be used. These should be strategies for both mitigation and adaptation — two words that are often poorly understood.

Student answers

Student A

Adaptation means coping with global warming's impacts, whereas mitigation means attempting to prevent global warming happening. **a** To mitigate, fossil fuel use needs to be reduced to lower carbon dioxide emissions. **b** This could be achieved by increased use of renewable energy such as the Thanet Wind Farm. **c** Adaptations could include planting drought-resistant crops **d** or strategic retreats as sea levels rise, e.g. the Blackwater Estuary in Essex **e** where sea walls have been removed because they are too costly to maintain as sea levels rise.

ⓔ **5/5 marks.** Grade A awarded. **a** The answer begins with a clear definition, **b** followed by a mitigation strategy **c** supported by a named example. **d** Two examples of adaptation are given **e** with a good explanation for the second. This answer uses examples and is well balanced.

Student B

Mitigation and adaptation are similar ways of tackling global warming. **a** You can mitigate by reducing pollution such as using solar panels **b** instead of coal or using wind turbines instead of gas. **c** People can mitigate by using cars that need less petrol, e.g. the Toyota Prius, so carbon emissions are lower. **d**

e **3/5 marks.** Grade C awarded. **a** The first statement is wrong, and suggests the student is not clear about the meaning of adaptation and mitigation. **b c** Two examples of mitigation are given although they are similar, plus the last example **d** which is more detailed and makes reference to carbon emissions. The answer has examples but is unbalanced.

Question 4

e These 5-mark questions, which usually appear at the end of a Section A question, often have a 'sting in the tail'. They could:

- use a phrase such as 'costs and benefits' or 'advantages and disadvantages', which requires you to cover both to write a balanced answer and achieve all of the marks
- have a very specific key word such as 'source' or 'host' countries, 'ecological' or 'natural hazards' — if you do not write about the key words in the question your mark will be low
- state 'using examples', which means you should refer to specific named places, players, organisations or schemes. If you don't refer to examples, you will not get full marks

On the move

Explain the costs and benefits of recent eastern European migration for source countries. (5 marks)

e Answers need to cover costs and benefits, but only for the origin countries (sources) of migrants not the destination countries (hosts). The migration type is post-2004 A8 migration from countries such as Poland.

Student answers

Student A

Countries such as Poland have lost up to 1 million workers due to emigration to the UK, **a** possibly meaning worker shortages in Poland due to the Brain Drain effect. **b** Other costs include an ageing population in Poland if workers stay abroad. Benefits include over 6 billion euros sent home in remittances in 2006 and a falling unemployment rate in Poland. Returning migrants could bring back new skills and ideas. **c**

e **5/5 marks.** Grade A awarded. **a** This is a good answer as it has factual detail about a named A8 country and it shows both **b** costs and **c** benefits, so is balanced. Two benefits plus two costs supported by examples and factual detail ensures maximum marks.

Student B

When people leave they cause a brain drain due to labour leaving the country. **a** Also, there could be fewer people left behind to fill jobs as well as pay taxes, e.g. for benefits. **b**

e **2/5 marks.** Grade D awarded. This answer is very general, and makes no reference to an A8 source country. **a b** Two costs are outlined; the point about taxes and benefits could have been extended by referring to dependency ratios. This is an unbalanced answer.

Section B
Question 5

The impacts of global warming

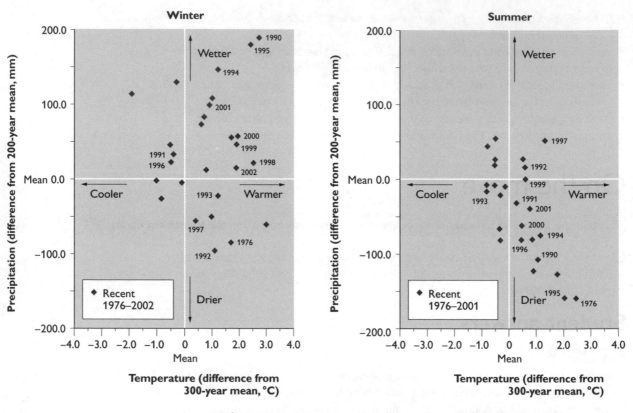

Figure 3 Variation in precipitation and temperature compared with the long-term means, 1976–2002, UK

(a) Study Figure 3. Analyse the patterns shown in the winter and summer variations in precipitation and temperature in the UK.

(10 marks)

 Your answer should focus on Figure 3, as this is a data stimulus question. You should refer to precipitation and temperature, quote data from the Figure and identify patterns (analyse).

(b) With reference to the Arctic, examine the environmental impacts of global warming.

(15 marks)

ⓔ Refer in detail to the Arctic, one of the compulsory case studies, but not other locations. Be careful not to drift from environmental impacts into social and economic ones.

Student answers
Mark scheme part (a)

Winters are in general getting wetter (18 are above the mean) with only nine drier than the mean. They are also getting warmer (only seven not warmer) but there are high variations.

Summers are less variable but there is a definite trend towards drier (all except 8) and also warmer (all except 11) summers. These changes are almost certainly linked to global warming.

Level	Mark	Descriptor
Level 3	8–10	An accurate analysis, with good use of detail. Shows good understanding of resources, well structured, recognises link to global warming.
Level 2	5–7	Some analysis of the data showing satisfactory understanding of the graph. Begins to describe pattern with some structure.
Level 1	1–4	Attempts to analyse patterns, but shows limited understanding.

Student A

(a) In winter, the graph shows that many **a** of the temperatures are warmer than the 300 year mean, and that in quite a few years it seems to be wetter especially in 1990 **b** which was the warmest and wettest year. In summer the pattern is not very clear to see for the temperature with some years a little warmer and some a little colder. However it does seem to have got much drier in recent years, with a lot of warmer and drier summers. The changes shown are very much the result of climate change which has led to rising temperatures. **c**

ⓔ **5/10 marks.** Grade D awarded. The graph has been understood and correctly interpreted. However, the answer is far too brief and it does not really make use of or analyse the data. **a** Numbers of years are not given but vague terms like 'many' are used, **b** and the date of only one extreme is given. It would have been useful to mention how much some individual years deviated from the mean. For example, 1976 had summers that were 2.5°C warmer than the 300-year mean, and over 150 mm drier. **c** The comment on climate change could be expanded to consider the impacts of global warming and the reasons for linkage to wetter winters.

Student B

(a) Figure 3 shows that in winter precipitation **a** has increased with many years in the 1990s more than 100 mm wetter than the 300-year mean, e.g. 1990, 1994 and 1995. **b** These years are also warmer than the mean by 1–4°C. **c** Thirteen of the years 1976–2001 are in this quarter of the graph, suggesting the overall trend towards warmer and wetter winters. **d** Colder, drier winters are very rare with only 3, and then only just.

    a For summer, most years are in the warmer and drier sector i.e. 12 years between 1976 and 2001 are in this area. There are 11 cooler than the average but they are only up to about 1.0°C cooler, so not really very different from the average. There are some extremes such as 1976. Overall the summer pattern is more clustered than winter so the pattern of warmer and drier summers is clearer than that for warmer wetter winters. **e**

ⓔ **10/10 marks. Grade** A awarded. **a** This answer is organised into two clear sections which cover winter and summer — this leads to a good structure. **b** This is a very analytical answer, which makes good use of data. **c d** The axes of the graphs are used to compare the mean and the sectors/quarters of the graphs are referred to, and the number of years in each is counted by way of analysis. **e** There is a useful overall summary, which makes the valid point that the summer pattern is more definitive than that for winter.

Mark scheme part (b)

This is a compulsory case study so examiners expect good detail on the precise changes — what is going on in terms of melting sea ice, rising temperatures, etc. They are looking for specific impacts on the ecology and the environment such as melting permafrost, poleward migration of vegetation belts, and impact on land and marine animals such as seals and polar bears.

Level	Mark	Descriptor
Level 4	13–15	Wide-ranging, well-structured account. A detailed examination of a range of environmental impacts. Well related to Arctic global warming. Well written.
Level 3	9–12	A sound account with a range of environmental impacts described. Structured response showing good links to Arctic environment.
Level 2	5–8	A general account, with some descriptions of several impacts. Geographical terminology infrequently used, sometimes in error. Some structure.
Level 1	1–4	Basic description of one or two impacts, poorly structured, with frequent language errors.

Student A

(b) There are numerous environmental impacts of global warming in the Arctic. Increasing temperatures have led to the coniferous forest belt moving north to where the tundra used to be. The result is that tundra ecosystems are being lost. **a** The increase of forests increases the risk of forest fires. **b** Also, the warmer temperatures allow the spread of diseases such as the spruce bark beetle. **c** The changes in the Arctic have had a huge impact on animals such as seals and polar bears who, because of the melting ice, have lost their habitat. **d** Also the rising temperatures have led to the melting of the permafrost, which has released a lot of methane into the atmosphere. **e** As many native peoples rely on the Arctic ecosystems for their traditional way of life the lack of species to hunt is affecting their way of life. The rising temperatures are also having an impact on the oceans, because rivers are bringing a lot of extra fresh water into the Arctic Ocean. **f**

ⓔ **8/15 marks.** Grade C awarded. **a** Some environmental impacts are covered, **b c** but they tend to be more of an extended list than a detailed discussion. **d** Inevitably polar bears are mentioned, but food chains and polar bear feeding are not — a missed opportunity. **e** The positive feedback caused by gas release from permafrost is missed. **f** Impacts on the oceans are covered in a generalised way. Overall, this answer has the skeleton of a very good answer, i.e. a good range of specifically environmental impacts, but without the depth required for a higher mark.

Student B

(b) The Arctic is probably the most sensitive environment in the world to global warming. The IPCC predicts the circumpolar region could warm by over 8°C by 2100, which would alter the place forever. **a** The most well known change is the melting of sea ice, which shrank 50%, in the summer, between 1979 and 2007. **b** Melting sea ice leads to two other changes:

1 The albedo goes from reflective to absorbing heat (the dark ocean), which will speed up warming via a positive feedback loop.

2 Polar bears can't hunt because the sea ice is how they get around to look for seals.

If polar bear numbers decline there will be knock on effects to the marine food web, with potentially more seals (the prey) and possibly fewer fish due to more seals eating them. **c** This could be a problem. On the other hand, as the Arctic Ocean warms there seems to be more cod in the seas, which is an environmental change with a positive human benefit for fishermen up north. **d**

The biggest change in the Arctic is likely to be a poleward shift in vegetation zones. **e** Warmer temperatures will push the taiga forest north, squeezing the fragile tundra into a narrower strip just south of the Arctic Ocean. This will have an effect on grazing migrants such as caribous, because their summer grazing lands will shrink. Even the expanding forests will not be impact free. In Alaska the spruce bark beetle **f** has spread into forests because winters are no longer cold enough to kill the beetle larvae; in the warmer summers forest fires are getting ever more common, destroying huge areas of forest.

ⓔ **13/15 marks.** Grade A awarded. **a** There is good detail, and a good reference to the IPCC's predictions of the scale of the warming. **b** Hard facts about sea ice, followed by two linked changes and good terminology (albedo, feedback). **c** Reference to food webs and how they could change is good, with some species referred to. **d** This answer recognises some positive impacts of environmental change, which gives the answer balance. **e** This is a very precise statement, and two Arctic ecosystems (taiga and tundra) are then named, with linked impacts to caribou. **f** Useful details are given on impacts in forest areas. Overall, this answer has both positive and negative impacts, and a wide range of environmental impacts and examples. For full marks it could have mentioned some Arctic locations in more detail.

Question 6

Globalisation

Figure 4 The global network of steel production and trade flows in 2006 (data in millions of tonnes)

Study Figure 4, which shows the global network of steel production and trade flows in 2006, and answer the following questions.

(a) Explain the uneven pattern of global steel production and trade flows. **(10 marks)**

> ℮ The key word is 'uneven'; think about why some places have a lot of production/trade in steel and some areas very little. Both production and trade flows must be explained.

(b) Using examples, examine the role of technology in the spread of globalisation. **(15 marks)**

> ℮ You should name places and use examples of technology, and explain how technology has helped globalisation (its role). Technology is a broad term that can include transport, communication and production technology.

Edexcel AS Geography

Student answers

Mark scheme part (a)

Trade is a fraction of production in all regions. This suggests transporting steel may be very expensive. There is some evidence of export to regions that are geographically close, e.g. Europe and Russia to the Middle East. Again, this could reflect high transport costs. Some regions that lack industrialisation are very dependent on imports, i.e. the Middle East. Production is heavily concentrated in Asia, with China (423 million tonnes) as the largest producer, around double any other. China is a low-cost NIC producer (global shift). China is a net exporter to higher-cost locations, e.g. Europe and the USA, where production costs are higher. North America is a large producer, but also a large net importer as production costs are lower elsewhere. This might suggest that production is still close to the TNC home country. Steel production and imports are lowest in the least developed regions, i.e. Africa, as there is little demand for it. You could also discuss the presence or absence of iron ore in the different locations.

Level	Mark	Descriptor
Level 3	8–10	Detailed explanations for flows and some patterns; shows good understanding of global network concept. Uses examples and appropriate geographical terminology to show understanding. Minimal written language errors.
Level 2	5–7	Some explanations for some patterns and flows. Becomes descriptive at the lower end but with some geographical terminology used. Some written language errors. Becomes descriptive at the lower end but with some geographical terminology used. Some written language errors.
Level 1	1–4	Little structure. Describes some aspects of the map without any real reference to geographical context or use of terminology. Frequent written language errors.

Student A

(a) The pattern is that production is concentrated in certain areas of the world. The biggest producers are North America, Europe, Japan and China. They produce about 200–300 million tonnes each year. **a** These are rich countries which control global steel production. However, places like China are industrialising and developing and need steel. **b** Some TNCs might have moved steel production to places like China because production costs are lower than in, say, the UK for instance. **c** Africa is a very poor country and has little need for steel due to most people living in slums so it has very small steel production and flows. **d** The flows are complicated with some places exporting and others importing. Russia is a big exporter and so is Latin America. North America is a big importer. As said previously, Africa is not really involved in production and trade. Some of the pattern could be due to trade blocks, for instance, there could be agreements allowing countries in Asia to import and export with no barriers, quotas and tariffs. **e**

e **6/10 marks.** Grade C awarded. **a** The data need to be used more precisely and more often to support the explanations of specific patterns. **b** This is a fair explanation, which could be developed. **c** Another fair explanation. **d** Errors such as the classic 'country of Africa' must be avoided. **e** The answer become descriptive at this point as the student struggles with the pattern on Figure 4. Trade flows are covered only generally and there is limited explanation and vague description. Mention of a specific trade bloc (not block) such as ASEAN would add detail and realism. Terms such as 'global shift' need to be included to show precise understanding. Overall, this answer has some explanations but they lack detail and development.

Student B

(a) Figure 4 shows the global network of steel making and trade in steel, mostly by ship. **a** One obvious variation is that the flow numbers are much lower than the production numbers, e.g. 9 from Europe to North America but 130 and 234 in North America and Europe. **b** This is because steel is very heavy and costly to transport so it is probably made where it is used if possible. **c** Large amounts of steel are made in the developed world, e.g. Europe (234), North America (130) and Japan (202). This is because it is used in industry such as cars, e.g. Nissan in Sunderland. **d** Even more, in fact the most, is made in China this is because China is rapidly industrialising and needs steel for skyscrapers, railways and its factories. Because it is made low cost, China exports steel. **e** A major flow is into the Middle East this could be for use in the oil industry, e.g. rigs, or for buildings in fast-growing cities such as Dubai. **f** Little steel is made in Africa and it has few flows. As the least developed area it has not a great need for steel as there are few factories and limited railways, ports etc. — at least developed ones.

ⓔ 10/10 marks. Grade A awarded. **a** This answer uses a word from the figure title, suggesting the student has studied it. **b** This obvious point could easily be missed; it is explained in terms of transport costs and data are used in support. **c** Different global regions are explained, again using data. **d e** These examples are good — China's production and trade are both explained in a clear way. **f** One of the major import areas is identified and an explanation for the imports is provided. The explanation for the Africa data is basic but perfectly acceptable. Overall, this is a good answer that covers both trade and production and refers to much of the map; explanations are clear and data are used. Quality of written communication is not perfect, but is readable and clear.

Mark scheme part (b)

The focus of your answer should be on the role of technology, although some drift away from the role of technology into other factors would be acceptable as the question does say 'examine'. Good answers might include the following key points and are likely to come to a brief conclusion.

The internet is an obvious example. It allows companies to communicate easily and cheaply with branch plants, customers and suppliers and plays a key role in allowing outsourcing. Individuals can 'explore' holiday destinations online. Mobile phones are similar in terms of 'keeping in touch' with widely dispersed networks. Air travel has shrunk the world (e.g. 747, Concorde, A380) since the 1950s, with budget airlines bringing the 'shrinkage' to the masses; this has allowed cultural mixing and migration as never before — it also aids the globalisation of TNCs. Containerisation is responsible for a revolution in transport costs and is a key factor in allowing TNCs to shift production globally. Bar code and other tracking software and devices allow complex companies like airlines and parcel carriers to operate globally.

Level	Mark	Descriptor
Level 4	13–15	Structured, detailed account that focuses on the role of technology. Uses appropriate terms and range of examples to show understanding. Likely to be evaluative in style. Written language errors are rare.
Level 3	9–12	Structured and uses some examples to explain the role of technology with appropriate geographical terminology to show understanding. Minimal written language errors.
Level 2	5–8	Some structure. Descriptive account which has some explanation of the role of some technology; likely to have narrow focus, but with some geographical terminology used. Some written language errors evident.
Level 1	1–4	Little structure. Describes one or two technologies with no real reference to globalisation context or use of terminology. Frequent written language errors.

Student A

(b) Technology has played a important role in globalisation and will continue to do so. The internet was invented in the 1980s and was beginning to be used commonly by the mid-1990s. **a** Today up to 1 billion people use the internet globally. The internet allows companies to communicate with employees and suppliers all over the world and this encourages companies to locate abroad (called 'outsourcing'), knowing they can still easily contact people. **b** The internet has allowed many TNCs to locate in the free trade zones in China and keep in touch with them. The internet is costly though, so not everyone in the world has it. **c** In poorer developing countries they may not have satellite and telephone networks, internet service providers, computers etc. Satellites are very important as these allow the internet to work, but most are controlled by TNCs or developed world governments which might explain why the internet has not spread to many very poor countries. **d** This means the internet is not available and this will discourage globalisation in these places. Africa needs the internet if it is to develop and this is why the '$100 laptop' has been introduced. The internet is very important in globalisation. TNCs are the next most important technology as some are more powerful than countries. **e** General Motors is as rich in turnover as Bulgaria. This gives TNCs the power to move production around the world and they play an important role in spreading globalisation.

ⓔ **9/15 marks.** Grade C awarded. **a** The internet is a good place to start, and there are some details of time scales and spread. **b** There are some explanations of how the internet has helped globalisation, e.g. outsourcing and TNCs. **c** The point about the internet not being universal is valid. **d** Other technologies such as satellites are mentioned, but the answer is starting to explain why technology is not found in some places, which is not the question. **e** TNCs are not a technology; the section on TNCs is not irrelevant, but it needs to be worded more carefully. Some appropriate terminology is used, such as outsourcing, but it is not used with enough precision. Some examples are included, such as the $100 laptop, but more depth and detail are needed. Overall, this answer is unbalanced. It focuses too heavily on the internet; reference to a range of technology is required to earn a higher grade.

Student B

(b) Globalisation is the increase in the number of global connections which has led to a shrinking world and increased flows of trade, money and people. **a** Lots of factors have helped globalisation but a key one is technology.

Communication technology has made it seem as if space has shrunk. **b** This started with telegraph and telephones more than 100 years ago, but accelerated with the invention of satellite communications in the 60s. Recently mobile phones and the internet have made communication instant worldwide. **c** Modern communications allow TNCs to be footloose, locating branches anywhere. **d** Apple Inc, based in the USA, has its products made by 3rd party suppliers in China **e** — without modern communications to send orders and make changes to design this outsourcing would be impossible.

Transport technology has also allowed world trade to grow so products can be got from anywhere in the world. Jet aircraft are the most obvious modern transport but containers are more important. **f** They were invented in the 1950s. Containers are intermodal steel boxes that can be cheaply and quickly transferred between ships, lorries and trains. They have reduced transport costs by over 80% and without them cheap goods from China, Taiwan and India would not be possible **g** — they are major drivers of world trade and there are over 17 million containers in the world.

People are more globalised than ever and are aware of distant places and can visit them. **h** This is because of technology such as:

- Social media e.g. Facebook and Skype keeping families and friends in touch.
- Global media and TV, which advertises far-off places.
- Cheap air travel, which means people can visit places which their parents could not.

This has created a 'global village' of connected people.

🅮 **15/15 marks. Grade** A awarded. **a** This is a good way to start, with a clear definition of globalisation. **b** The type of technology to be discussed is stated clearly. **c** The discussion of communication technology over time is good, and concise. **d** There is a good link to how global companies benefit from modern communications. **e** It is linked to a good example of a TNC. **f** Transport technology is discussed in detail and the depth of knowledge about containers is impressive, **g** especially the explanation of reduced transport costs. **h** This is an interesting section about people and how technology has affected them. Overall, a range of technologies are outlined in some detail, using examples and the 'language' of globalisation and global connections. The answer is structured into communications, transport and people sections.

Question 7

Global hazard patterns

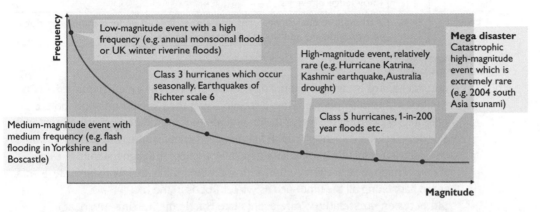

Figure 5 The relationship between natural hazard magnitude and frequency

(a) Study Figure 5. Examine the relationship between hazard magnitude and frequency.   (10 marks)

ⓔ Be clear about the meaning of magnitude and frequency. Answers need to focus on the relationship, i.e. link between magnitude and frequency, and use examples to illustrate it.

(b) Explain how human factors such as deforestation and urbanisation can increase the risk of natural disasters.   (15 marks)

ⓔ This part of the question focuses on natural disasters, not natural hazards as in part (a); notice that both deforestation and urbanisation are mentioned, but the 'such as' means you can refer to other human factors.

Student answers

Mark scheme part (a)

An inverse relationship is shown: higher frequency = lower magnitude. Quite small increases in magnitude correlate with very significant drops in frequency for lower values.

A variety of different hazard types can be explored.

Plenty of well-known examples are provided as a stimulus for students to add their own ideas and information. There are other important recent events such as the 2009 Haiti earthquake or 2010 Pakistan floods that could be used as part of the answer.

Return periods ('1-in-200') are used but not explained in Figure 5 and good answers may be able to explain how they are used in hazards analysis.

Level	Mark	Descriptor
Level 3	8–10	An accurate analysis, with good use of detail. Shows good understanding of the resource, well structured, recognises that a wide range of hazards are subject to the same statistical trend.
Level 2	5–7	Some analysis of the data showing satisfactory understanding of the graph. Describes the correlation with some examples that may draw on own knowledge of events.
Level 1	1–4	Attempts to analyse patterns, but shows limited understanding and does not provide any additional information beyond that given.

Student A

(a) A natural hazard is caused when a physical event such as an earthquake has an impact on vulnerable people. This is shown using the disaster risk equation. **a**

Figure 5 shows that the bigger the magnitude the less the frequency is for instance mega-disasters are catastrophic high magnitude events and they are extremely rare such as the tsunami that hit South Asia in 2004. **b** These are high magnitude events that do not happen very often. At the other end of the scale are low magnitude events that are happening all the time such as winter floods in the UK.

In between, are relatively rare events like Hurricane Katrina. Hurricane Katrina struck New Orleans in 2005 there was not enough time to evacuate everybody and many lives were lost. People took shelter at the sports centre but the tragedy was made worst when it turned out that many of the poorest people were not insured or had their homes looted. This meant that the impacts of this relatively rare event were made even worse. **c**

So the relationship is that the more extreme an event is the more rare it happens and the impacts can be even worse when vulnerable people are exposed. **d**

ⓔ **4/10 marks.** Grade D awarded. **a** The student begins by identifying the relationship as asked, although it is not very well expressed. **b** Much of the answer is copied directly from the figure with nothing else added from the student's own knowledge. **c** Although the information about New Orleans does not appear in the figure, it does not help address the question and instead focuses on a different issue, namely people's vulnerability. **d** The same uncertainty about the meaning of the question reappears in the last sentence.

Student B

(a) Figure 5 demonstrates an inverse relationship between hazard magnitude and frequency. It is also non-linear (at low magnitudes a slight increase in strength can cause a large reduction in frequency). **a** An example of a high magnitude mega disaster is the 2004 Asian Tsunami that was triggered by an underwater earthquake off the coast of Indonesia and killed 280,000 people. Such events are extremely rare. However, every day there are hundreds of small under sea earthquakes that do not have sufficient magnitude to present a risk to people.

The flooding of Boscastle in 2004 was estimated to be a 1-in-400-year event. **b** This means that in any year there is only a 1-in-400 chance of it happening. A wall of water 10 feet high travelled through Boscastle — an extremely rare event. **c** However, every year there are many much smaller floods in the UK because statistically they are much more likely to happen. **d**

e 9/10 marks. Grade A awarded. **a** A clear statement of the relationship; this starts with a good focus with no time wasted. **b** Information from Figure 5 is used selectively along with the student's own knowledge of the Asian tsunami and Boscastle. **c** There is good understanding of return periods. **d** The summary shows a good grasp of the relationship. The answer could have made further reference to Figure 5 for maximum marks.

Mark scheme part (b)

Students are required to learn about the role of deforestation and urbanisation — as well as additional human factors like poverty and population growth.

Human factors can modify physical processes thereby increasing the hazard threat. Run-off from deforested or urbanised areas is a key cause of increased flooding and landslide risks.

Human factors are also an important element of the disaster risk equation as these factors (especially poverty) can influence people's vulnerability and their capacity to cope when a hazard strikes.

Level	Mark	Descriptor
Level 4	13–15	Wide-ranging, well-structured account. A detailed examination of a range of human factors including those specified in the question. Well related to natural disasters. Well written.
Level 3	9–12	A sound account with at least two human factors explained. Structured response showing good links to natural disasters.
Level 2	5–8	A general account, with some descriptions of one or more human factors. Geographical terminology infrequently used, sometimes in error. Some structure.
Level 1	1–4	Basic description of either deforestation or urbanisation, with frequent language errors.

Student A

(b) There are many different kinds of natural disaster. Geophysical hazards are caused by tectonic activity along plate boundaries and include earthquakes and volcanoes. Hydrometeorological hazards include floods, storms and droughts and are on the rise on account of climate change. **a**

Deforestation is one way in which humans can increase the risk of natural disasters such as flooding. When forest is removed it destroys the vital interception store that can slow down overland flow which is a major cause of river flooding. **b** In the Amazon rainforest millions of square kilometres of forest has been cut down by TNCs in order to create space for the growth of crops such as soya beans. The soya beans are needed in China, which is one of the BRICs and where rising wealth has meant there is a greater demand for food. **c** In Brazil, not enough has been done to prevent companies from stripping the forest away which has led to the Amazon river flooding during periods of heavy rainfall. **d**

Deforestation has also taken place in the Philippines, a disaster hotspot where many vulnerable people live. **e** The same problem has occurred, with the lack of forest cover meaning that when the hurricane rains arrive the water is not held by the forest's leaves and instead flooding occurs. Urbanisation also creates problems as the concrete in cities such as Manila is a major cause of flooding. **f**

ⓔ **8/15 marks.** Grade C awarded. **a** There is a lot of generality about types of hazard, which does not focus on the question. **b** Some terminology is used in the section on deforestation. **c** The answer begins to drift into a lot of detail about soya beans and food production, rather than focus on the causes of disasters. **d** Eventually the answer returns to flooding. **e** The section on the Philippines mentions vulnerability and links deforestation to flooding. **f** There is a hint of urbanisation, but this needs development. Overall the answer is a partial answer to the question, unbalanced by concentrating on deforestation.

Student B

(b) Natural disasters occur when physical events such as hurricanes or earthquakes strike a vulnerable population, resulting in loss of property or life. **a** Human factors are therefore a big part of the overall equation, especially people's vulnerability and their capacity to cope. **b** A range of human actions such as deforestation and urbanisation can also mean that the physical processes become harder to control and have greater effects, which is the focus of this essay. **c**

Urbanisation is a growth in urban areas and often takes place next to rivers or along coastlines, i.e. areas with a flood risk. In California, Los Angeles has developed along the San Gabriel river whose hillsides have also been deforested in recent years. Winter storms, especially during El Niño years, trigger heavy precipitation in the catchment. Concrete and tarmac are impermeable and do not allow water to infiltrate. The result is overland flow that quickly reaches the San Gabriel, leading to major flooding. **d**

The same problem caused by urbanisation was seen in the UK when Boscastle flooded in 2004, and Carlisle in 2005. With nowhere to go, rainwater quickly reached the river, making the flood disaster when it happened much worse. **e**

Turning now to deforestation, this is seen widely in the Amazon rainforest, African countries such as Mozambique and in the Philippines. In all cases, stripping away the canopy means that monsoon or hurricane rain can no longer be intercepted. **f** The problem is made worse when slopes are steep or soils are very thin.

Coastal flooding is also made worse by deforestation. Mangrove forest removal is one reason why the 2004 Asian tsunami had such disastrous effects in parts of Thailand and Sri Lanka, as nature's protection had been removed for agriculture and aquaculture. **g**

Urbanisation takes place in many locations and will always result in increased disaster risk if hazards are present, thus the growth of Tokyo (the world's largest city) on an active plate boundary has increased the risk from volcanoes and earthquakes. **h** Similarly the urbanisation of Los Angeles has increased the disaster risk associated with the San Andreas fault — the next 'big one' will cause massive financial losses. Finally, the human factor of climate change is thought to be a cause of increasing extreme weather events across the globe. **i**

ⓔ **13/15 marks.** Grade A awarded. **a** A useful start, with a clear definition of natural disaster. **b** Good use of technical terminology. **c** The introductory paragraph is good, as it sets up the detailed content to follow. **d** Urbanisation is defined, followed by a detailed example that cleverly links to deforestation and El Niño. This is a good explanation. **e** Further examples are used to reinforce the point. **f** Deforestation is explained, for several locations, using terminology. **g** A named disaster is directly linked to coastal deforestation. **h** Urbanisation is linked to rising risks from other hazards, although the nature of the risk could have been spelled out (increasing number of people in vulnerable places). **i** The comment on climate change could have been extended by the use of an example.

World at risk

1 A hazard is a natural event, which can be defined as a threat to humans and the environment. Disasters are the realisation of hazards to cause social and economic impacts such as deaths, and loss of livelihoods, goods and property, i.e. the actual consequences of the hazard on vulnerable people.

2 Positive, as the melting of ice releases large quantities of methane — a very powerful GHG — so amplifying global warming.

3 Thermal expansion of the oceans. Future increases in the rate of sea level rise will result from melting glaciers and ice sheets.

4 Kyoto 1997, Copenhagen 2009.

5 Hydro-meteorological — linked to short-term climate change.

6 Tsunamis, coastal flooding from storm surges, hurricanes.

7 Economic impacts are concerned with destruction of economy, infrastructure, loss of jobs, whereas social impacts can lead to death and disaster-induced stress from homelessness and famine, and health risks.

8 These are years of mega disasters: 2004/05 Boxing Day tsunami, Kashmir; 2009/10 Haiti earthquake.

9 Both increasing personal wealth and growing economies especially in the NICs lead to much higher value losses and insurance claims.

10 'Absolute terms' is the total value of economic losses — likely to be greatest in developed countries, whereas relative terms include the percentage of the total GNP lost — likely to be very high in LDCs.

11 (a) Earthquakes, (b) Volcanoes, (c) Drought.

12 Constructive — less of a jarring movement as plates are pulled apart and are not in collision.

13 Conservative — no magma is intruded at this boundary.

14 Mauna Loa and Kilauea (Hawaii). Tristan da Cunha.

15 Young tertiary fold mountains, frequently deforested, lead to slides after heavy typhoonal rain.

16 Drought is complex to define. It is associated with a long, continuous period with little or no precipitation, leading to an imbalance between natural water availability and human demands.

17 Interior continental areas are often far from rain-bearing winds and therefore receive low total and irregular rainfall, which makes them drought prone.

18 1. Concentrated between 5° and 20° N and S of equator. 2. Not at the equator itself. 3. Greatest concentration of strong hurricanes off East Asia. 4. Greater numbers in northern hemisphere.

19 Concentration of several types of hazards in an area with large concentrations of vulnerable people. This leads to frequent disasters from multiple hazards.

20 The area is sufficiently north of the equator to be located on the track of typhoons swinging in from the Pacific and moving northwest towards China.

21 Destructive boundary. Philippines plate is subducted beneath Eurasian plate.

22 Earthquake.

23 Earthquake.

24 Indirect substitute evidence concerning climate change from pictures, poetry etc., i.e. not directly measured.

25 High sunspot activity leads to warmer periods, e.g. medieval warm period.

26 Even with very large — eruptions, e.g. Krakatau — temperature changes only last for 1–2 years.

27 'Astronomic forcing' is when variations in the Earth's orbit cause a change in the amount of solar radiation, whereas 'atmospheric forcing' refers to the enhanced greenhouse effect.

28 The greenhouse effect is when more heat is trapped by the greenhouse gases and less is radiated back into space. Higher concentrations of GHG lead to an enhancement or increase in this effect and result in warming global temperatures.

29 The increase in Arctic temperatures permits conifer growth further north, i.e. in the present-day tundra zone.

30 The continent of Africa contains most of the world's poorest peoples. A combination of factors makes the continent highly vulnerable to even modest impacts of climate change.

31 East Africa will get more rainfall for farmers.

32 Rising sea levels result indirectly from temperature change, which leads to thermal expansion of the oceans.

33 An environmental refugee is a person who flees or is forced to migrate from an adverse environment in order to live. Char Islands are in a delta area and Tegua is a coral atoll — both are at the front line of sea level rise as they are so low lying.

34 Nobody can predict the level of GHG emissions, as it depends on the extent to which damaging actions can be mitigated.

35 There is a time lag in the system and it takes time for the effects of mitigation to 'kick in'.

36 Individual — conserving energy or fitting solar panels on roof. Local — developing sound recycling strategies, or encouraging bike use. National — developing a carbon tax to cut emissions, or taxing gas-guzzling cars. Global — agreeing target levels for cutting emissions.

37 Actions such as building coastal defences or improved water storage (climate proofing) only take 2–3 years to implement locally, whereas mitigation strategies take a very long time to have an impact.

38 Greener growth attempts to use more sustainable strategies to cut GHGs.

39 Examples: providing cycle lanes gets people out of their cars, cutting petrol consumption. Taxing city centre parking spaces forces people onto greener public transport. Local farmers markets cut down on food miles and fossil fuel-powered transport. LED lights conserve energy so cutting down on electricity use — most electricity is produced by fossil fuels.

40 Carbon footprints are restricted to calculations of the amount of CO_2 emitted, whereas eco-footprints cover a much wider area, e.g. use of all resources.

41 Rainfed agriculture relies directly on rainfall supply as opposed to irrigation.

42 'Physical water shortage' refers to a lack of availability of water, with demand exceeding supply, whereas 'economic shortage' means that water is available but people cannot afford it.

43 Biofuels are made from plant or animal matter, e.g. biodiesel, biogas, bioethanol.

44 They are eco friendly because the carbon they give off is balanced by the carbon captured in their production but some of the biomass used is planted in cleared rainforest areas, or it takes land used for food crops.

45 Intermediate technology is non hi-tech. It is cheap to buy and maintain — entirely appropriate for developing countries.

46 They are long awaited as they may help to feed the world's people, as second generation GM crops will be drought resistant and salt tolerant and suitable for marginal lands.

47 Bottom-up strategies involve the local people in the decision making and carrying out the plan. These grassroots initiatives are likely to be successful. NGOs etc. train local people to take the initiative.

48 Energy efficiency actually saves money, by using less energy. It is therefore both environmentally friendly and cost effective.

49 It was developed at the World Conference on Disaster Reduction as a blueprint for action — held in Hyogo, a district in Kobe in Japan.

50 Sustainable communities develop strategies for cutting GHG emissions and also plan to become climate proof by adapting to climate change. They also develop community-based strategies for managing disasters. The result is capacity building, so a community resilient to risk is developed using this integrated response.

Going global

51 Mobile telephones and the internet.

52 'Foreign direct investment' is where a TNC spends capital in another country setting up a factory, business, joint venture or infrastructure project.

53 Containerisation and jet air travel have speeded up the movement of people and goods, and reduced transport costs.

54 Most migration is driven by economic motives, specifically the search for a job and a better income. Most illegal immigrants work in the grey economy in order to survive.

55 NICs and RICs have a high percentage of jobs and economic activity in manufacturing. They are often major exporters of the goods we buy in shops, e.g. China.

56 The UK is a member of the EU and is also an OECD country. The UK is also a G8 and G20 member country.

57 Africa, and especially sub-Saharan Africa, has benefited the least from the growth of global trade in the last few decades.

58 Exxon Mobil, BP and Royal Dutch Shell are all major oil companies. Toyota Motors is a car and truck manufacturer and Walmart is a retailer.

59 Glocalisation means adapting global products and brands to suit local tastes and market conditions, as well as sourcing raw materials and components locally.

60 When TNCs leave a source country and relocate some activity overseas it may lead to unemployment in the source country, derelict land and abandoned factories, and often leads to negative publicity.

61 Increased air travel has connected people, but it is a major source of greenhouse gases that contribute to global warming.

62 Air freight goods tend to be small, light, relatively high value and/or perishable, as air freight is costly compared to shipping. They include cut flowers, small components, urgent mail and packages, and medical equipment.

63 A global hub is a location, usually a city, with many types of connections to other locations. These include communication and transport links, flows of people and economic connections such as TNC headquarters and branches. Global hubs are said to be 'hyper-connected'.

64 North America, Europe and Oceania are the most connected regions. The developing regions, especially Africa, which has close to 20% less internet usage than the world average, are the least connected. Asia is also below the world average.

65 The main losers are likely to be the colonial banana producers who sell expensive bananas. Winners include the dollar banana growing countries, the TNCs that dominate the trade and possibly EU consumers who may get cheaper bananas.

66 The UK has low fertility, below the replacement level of 2.1 and a low birth rate of 10.7/1000. Life expectancy is high and rising, and the average age of the population is now 39.

67 Deindustrialisation means the decline in the number of people employed in the secondary (manufacturing) sector as factories have closed. Most UK jobs are now in the service sector.

68 The UK population pyramid is top heavy, with a narrow base. Numbers of people aged 0–29 are lower than the numbers aged 30–54. The pyramid is tall, as there are significant numbers aged over 80, especially females.

69 Governments in the future are likely to face increased bills for old age pensions, NHS care for the elderly, and for care and nursing homes.

70 It was government policy to encourage migrants from Commonwealth countries to fill a skills shortage, especially in low-paid jobs. Strict immigration laws did not exist in the 1950s and 1960s, which encouraged family members to migrate to join earlier migration. Also, the UK was familiar to migrants, many of whom spoke English.

71 Immigrants increase population numbers directly when they arrive, and often indirectly via higher birth rates and/or by family members joining them.

72 The A8 countries are Poland, Latvia, Lithuania, Estonia, Czech Republic, Slovakia, Hungary and Slovenia; they joined the EU in 2004.

73 A8 countries have benefited from worker remittances, migrants may take skills back in the future, and migrants working abroad do not need benefits in their home country. Skills shortages, brain drain and lack of economic innovation are all costs of losing workers to the UK.

74 Migration is made easier by the free movement of people within the EU, the growth of low-cost airlines, electronic communication, and companies that sort out the paperwork for migrants.

75 A combination of factors including perceived relaxed lifestyle, the climate, lower costs of living, attractive new villas, other Britons already there and the relative ease of the migration process.

76 2007–08 was the first year when more than 50% of the world's population was living in urban areas.

77 A megacity is any urban area with a population of over 10 million. To be a world city the urban area must have global economic and/or political significance.

78 Urbanisation results from the twin processes of rural–urban migration and internal growth (the birth rate in urban areas).

79 Segregation means that the rich and poor live separate lives, in separate areas — the very poor in slums and the rich in gated communities.

80 They aim to reduce the ecological footprints of urban dwellers by minimising inputs (energy use) and outputs (waste) as well as providing people with a high, but sustainable, quality of life.

81 If people own the land their home is built upon they, and their wider community, have an incentive to improve their home and area.

82 Food miles matter because transport of food over long distances uses fuel, which releases carbon dioxide, contributing to global warming.

83 Factory workers have moved there because the economic prospects are better than in rural China, as wages are higher.

84 Small cars with small petrol engines, diesel cars and hybrid or electric cars have low VED.

85 Fair trade gives a greater income to developing world farmers, by guaranteeing them a price for their produce.